Start Right in E-Business

**A Step-by-Step Guide
to Successful E-Business
Implementation**

D0226336

E-Business Solutions

An Academic Press Series

Bennet P. Lientz and Kathryn P. Rea
Series Editors

The list of titles in this series includes:

Start Right in E-Business
Dynamic E-Business Implementation Management
Breakthrough Technology Project Management, 2nd ed.
Grow Your E-Business for Success

Start Right in E-Business

A Step-by-Step Guide to Successful E-Business Implementation

Bennet P. Lientz

Information Systems
Anderson Graduate School of Management
University of California, Los Angeles, California

Kathryn P. Rea

The Consulting Edge, Inc.
Los Angeles, California

ACADEMIC PRESS
A Harcourt Science and Technology Company

San Diego   San Francisco   New York   Boston   London   Sydney   Tokyo

Academic Press
A Harcourt Science and Technology Company
525 B Street, Suite 1900, San Diego, California 92101-4495, USA
http://www.academicpress.com

Academic Press
Harcourt Place, 32 Jamestown Road, London NW1 7BY, UK
http://www.academicpress.com

Library of Congress Catalog Card Number: 00-107118

International Standard Book Number: 0-12-449977-5 (pb)

PRINTED IN THE UNITED STATES OF AMERICA
00 01 02 03 04 05 ML 9 8 7 6 5 4 3 2 1

Contents

Part I
Get Started in E-Business

Chapter 1
Introduction

Chapter 2

Your E-Business Implementation Roadmap

Part II
Collect Information on the Business Processes and Technology

Chapter 3
Action 1: Understand Your Business

Chapter 4
Action 2: Select the Activities for E-Business

Chapter 5
Action 3: Assess E-Business Trends and Competition

Chapter 6

Action 4: Set Your Technology Direction for E-Business

Chapter 7

Action 5: Collect Information for E-Business

Part III

Define How E-Business Will Work for You

Chapter 8

Action 6: Analyze Information for E-Business

Chapter 9
Action 7: Define Your New E-Business Transactions and Workflow

Part IV
Prepare for Your E-Business Implementation

Chapter 10
Action 8: Define and Measure E-Business Success

Chapter 11

Action 9: Develop Your E-Business Implementation Strategy

Chapter 12
Action 10: Perform Your E-Business Marketing Activities

Part V
Implement Your E-Business

Chapter 13
Action 11: Plan Your E-Business Implementation

Chapter 14
Action 12: Execute Your E-Business Implementation

Chapter 15
Action 13: Follow Up
after Your E-Business Implementation

Part VI
Address E-Business Issues

Chapter 16
E-Business Implementation and Operations Outsourcing

Chapter 17
Address Specific E-Business Implementation Issues

Preface

E-BUSINESS AND E-COMMERCE

Electronic business, or E-Business, is a hot topic. One estimate is that it will have grown from $10 billion in 1997 to over $200 billion by 2001. Another firm estimated that in five years its volume will be greater than the gross domestic product of Great Britain or Italy. It is thought that within 7 years over half of American businesses will be doing E-Business. E-Business through e-commerce transactions will transform businesses and relationships between customers, suppliers, products, services, and marketplaces.

Of the many firms that attempt to implement E-Business, a high percentage either fail or at best only partially succeed. What are some of the reasons for failure?

- The companies do not take an appropriate approach to E-Business. They often treat it as just another IT project. E-Business is not that—it is very different.
- E-commerce transactions are treated as totally separate from the regular stream of business. Lack of coordination and support can doom your E-Business efforts.
- Organizations aim for quick and dirty E-Business implementations on top of their existing systems and business. The older business models and technologies do not fit into E-Business and customers leave in droves.
- Companies underestimate the initial effort of implementation and ignore the ongoing demands of the marketplace. Complacency leads to loss in market-share.
- E-Business efforts are often too narrowly focused. A Web site with a catalog and online ordering are viewed as sufficient.

In fact, part of the problem is the attention to merely setting up some of the basic transactions of the business for electronic transactions. Our focus is the transformation and change of the enterprise to electronic business or E-Business while simultaneously improving internal business methods. This is the core of "starting right in E-Business." This is more ambitious than "e-engineering," which attempts to more narrowly establish electronic commerce. Experience shows that if you just focus on the "e" part of the business and not the rest, you have a greater risk of failure. That is what is behind our title—*Start Right in E-Business*. We want you to learn from past problems and avoid them.

Let's stick to the basics and common sense. E-Business occurs when you have established critical business procedures and activities to support e-commerce transactions. E-commerce is then part of E-Business using this definition. You need e-commerce to implement E-Business. However, and this is important, just because you have implemented e-commerce, you have not necessarily implemented E-Business. You only implement E-Business when you have changed business procedures to take advantage of the e-commerce technologies. You have implemented E-Business when you have also changed your internal methods to take advantage of the "e."

E-BUSINESS IMPLEMENTATION

Implementing E-Business requires improving and modifying how you do business—not only initial electronic business, but also continuing change. There is a sharp contrast between E-Business implementation and reengineering. In reengineering you attempt to make major changes in business processes with the vague goals of cost reduction and productivity. For many firms, reengineering has meant downsizing. It is not surprising that the term "reengineering" has generated such strong negative feelings. E-Business implementation has a specific positive goal—to compete more successfully in the modern world using electronic trading, commerce, and transactions.

E-Business implementation has a directed focus as well. In reengineering you are free to select any activity to analyze and change. E-Business implementation forces you to consider all work that is involved in any way in the transactions that will be electronic. Thus, if you are going to sell products on the Web, you must include ordering, customer service, order fulfillment, shipping, and accounting. You cannot stop with ordering. An example will highlight this. A firm went on the Web with a very good Web site to sell books, CDs, and magazines. The ordering approach was extremely efficient. However, if you did not receive the book, you could not obtain status information on the Web. You had to call a toll-free number and talk to an employee of the firm. This employee manually wrote the information down and faxed it to the warehouse. The warehouse located the order and faxed back a response to customer service. The customer service employee then called the customer. The cost of the manual steps consumed all of the profit in the transaction. Moreover, customer service was abysmal. The firm later closed its doors.

E-BUSINESS STRATEGIES

Another factor impacting E-Business implementation is your strategy for E-Business. There are several choices. You can proceed with E-Business as a totally separate activity. One reason for this approach is to create a different image for your E-Business. Another strategy is to piggyback E-Business on top of regular business. A third strategy is to integrate E-Business and business. Finally, you can change some of your current business activities into E-Business ones. You should consider all four of these potential E-Business strategies. All of these are taken up in the book. However, even if you establish E-Business separately, you still must coordinate the direction of change of both your regular business and your E-Business.

OBJECTIVES OF THE BOOK

We want your organization to be successful in implementing E-Business in the 21st century. What is success in E-Business? Here are the goals that you should shoot for:

- Your regular and E-Business activities integrate and interface almost seamlessly.
- Through implementing your E-Business strategy you refresh and renew your existing business.
- Your E-Business implementation provides you with flexibility to respond to new market opportunities and challenges from industry as well as new technologies.
- Your employees are motivated and supportive of your E-Business initiative. They feel that they are part of the action.
- Your implementation of new methods and approaches can sustain and increase your competitive position and advantage.

There are basic principles and factors to keep in mind. Competitive advantage is both difficult to achieve and even more difficult to sustain. Why? Solutions in E-Business can be copied. Thus, to be successful you must have a dynamic approach to all of your critical business activities and business strategies.

WHO CAN BENEFIT FROM THIS BOOK?

Our target audience consists of managers, employees, and consultants who either are involved or seek to become involved in the dynamic world of E-Business. Since the approach includes automated systems, organization, business methods, and policies, it is comprehensive. The approach does not assume that you have a technical background or that you have experience in E-Business or e-commerce. More specific audiences include:

- Managers and staff who are chosen to head up E-Business implementation efforts.
- Managers who are considering E-Business implementation and want to know the best approach to use.
- Employees who are involved in an E-Business implementation.
- Entrepreneurs who are interested in starting up an E-Business.
- Consultants who want to pursue E-Business implementation as a line of business.
- IT professionals who want to establish themselves and their groups firmly in E-Business.

ORGANIZATION AND BENEFITS OF THE BOOK

This book contains a step-by-step approach to E-Business implementation. E-Business has been implemented in a number of organizations successfully. It has been used in different countries and industries and in companies of widely different sizes. Each chapter provides detailed guidelines for helping you transform your business into an E-Business enterprise. To make this even more useful to you, over 200 guidelines and lessons learned from experience in over 50 E-Business implementation projects have been gathered and included.

People who have employed this material have pointed to the following benefits:

- They can implement E-Business in *less time.*
- Implementation of E-Business comes with *reduced risk* to the existing business.
- E-Business is implemented in a *coordinated manner* with current business activities.
- Overall, the approach is *less costly* than some methods.
- The method gets employees and managers *more motivated* to implement E-Business. They feel that they have more "skin in the game."
- The approach is *more comprehensive* than just e-commerce.
- *More flexible* business procedures tend to result.

The book is divided into logical parts that begin with getting you started in E-Business and continue through reviewing your results after implementation. In addition, there are chapters that address specific issues and opportunities, such as outsourcing, that you are likely to encounter in implementing E-Business. We have purposely not used the word "process." In E-Business some of the things that you will implement and change are not processes. They are activities such as marketing and data analysis. Thus, we have used the phrase "business activity" to be more general.

About the Authors

Bennet P. Lientz is a consultant, teacher, and researcher in E-Business. He has advised startup E-Business firms as well as helped firms move into E-Business. He is Professor of Information Systems at the Anderson Graduate School of Management, University of California, Los Angeles (UCLA). Dr. Lientz was previously Associate Professor of Engineering at the University of Southern California and department manager at System Development Corporation, where he was one of the project leaders involved in the development of ARPANET, the precursor of the Internet. He managed administrative systems at UCLA and has managed projects and served as a consultant to companies and government agencies since the late 1970s.

Lientz has taught project management, information technology, and strategic planning for the past 20 years. He has created two E-Business courses. He has delivered seminars related to these topics to more than 4000 people in Asia, Latin America, Europe, Australia, and North America. He is the author of more than 120 books and 60 articles in information systems, planning, project management, and E-Business.

Kathryn P. Rea is president and founder of The Consulting Edge, Inc., which was established in 1984. The firm specializes in E-Business, information technology, project management, and financial consulting. She has consulted with over 45 organizations in E-Business implementation and expansion.

Rea has managed more than 65 major technology-related projects internationally. She has advised on and carried out projects in government, energy, banking and finance, distribution, trading, retailing, transportation, mining, manufacturing, and utilities. She has successfully directed multinational projects in China, North and South America, Southeast Asia, Europe, and Australia. She has conducted more than 120 seminars around the world. She is the author of eight books and more than 20 articles in various areas of information systems and analysis.

Part I

Get Started in E-Business

Chapter 1

Introduction

E-BUSINESS CONCEPTS

A *business activity* is a set of procedures and workflow steps that carries out a specific business function. Payroll, ordering, inventory control, shipping, and fulfillment are examples of business activities. A business activity is composed of many discrete business transactions—some formal and some informal. Considering this broad definition, business activities encompass almost everything in the firm. The most important business activities are those that add to a company's competitive position and provide value to customers. It is these activities that E-Business implementation addresses. E-Business activities include not only the electronic commerce transactions with customers and suppliers, but also the supporting internal transactions—some of which may not be electronic.

When you do E-Business, you will likely employ a mixture of some of the following E-Business components:

- Internet or web site
- EDI (Electronic Data Interchange)
- Business to business sales over a supply or value chain
- Customer self-service
- Web-based e-commerce
- Call center integration
- Extranet for the supply chain network

Surveys indicate that firms have gotten a rapid payback for their E-Business investment if the underlying business activities were improved correctly. Firms cite as benefits the following:

- Improved and changed business model (generally improved procedures and simplified policies)
- Enhanced competitive position (more efficient business)
- Improved employee communications (pointing to collaboration and changed procedures)
- Improved customer service and satisfaction (improved effectiveness)
- Reduced costs (increased efficiency)
- Greater understanding of customers and markets (new procedures and workflow created)

E-Business implementation aims to implement electronic transactions as well as modifying current business activities to integrate and interface with the electronic transactions. Implementing E-Business then transforms your current business as well as enabling you to support electronic business.

At the heart of E-Business implementation you are installing new methods and procedures as well as changing current ones. In the past improvements have often employed continuous improvement or reengineering—with very mixed results. In continuous improvement you seek to change the company processes over time. Unfortunately, this is too slow for E-Business implementation. Reengineering aims at radical process change. This sounds nice, but many companies have found that this doesn't work. The organization becomes traumatized. People become resentful since they perceive that their jobs are threatened. It is no wonder that the success rate is less than 50%. In E-Business implementation, a key goal is to involve many employees—down to the lowest levels of the organization. Participation leads to commitment. Commitment leads to E-Business success.

Our approach has the following themes.

- It is critical to involve many employees in E-Business implementation.
- The current business activities cannot be disrupted since they are the mainstays of the survival of the business.
- You must take an integrated approach in terms of new and current business activities and transactions.
- To get E-Business going will be one large project. However, with tight deadlines and given the extent of the work involved, you will find that you must implement E-Business as several projects under an overall umbrella project. Competitive pressures mean that you will likely have E-Business projects going on all of the time.

Some standard terminology will be used throughout the book. Here are some definitions of frequently used terms:

- *E-commerce.* Business transactions conducted on-line.
- *E-Business.* Your business activities, organization, and structure that use e-commerce transactions.

- *Infrastructure.* Computers, communications, facilities, and any other physical support required by business transactions, including equipment, information systems, general technology, and other resources that support implementing E-Business and standard business procedures.
- *Organization.* The business's structure and its employees.
- *Activity group.* A group of business activities that are related through one facet of a business, such as organization, technology, suppliers, or customers.
- *Technology.* All categories of technology involving computers, software, and communications.

Why do businesses change how they do their work to implement E-Business? E-Business implementation can take place for both external and internal reasons. External reasons include changes in competition, supplier–customer relations, government regulation, and technology. Internal reasons may include management direction, lack of flexibility, and obsolescence of current work practices. Change is also motivated by the simple fact that the existing business activities do not live up to management expectations—on top of E-Business pressures. Perhaps, people have been focused on the "how" and not the "what," "who," or "where." Perhaps, the people performing the work feel comfortable and secure with "the way it has always been" and have not reevaluated the workflow for awhile. Whatever the reason, the current workflow and procedures are failing or have already failed and will not support E-Business as they are. Another basic premise of the book is that since you are spending all of this money and energy on E-Business, you should achieve flexibility and improve current workflow and procedures at the start—to the extent that time permits.

When change is called for, pouring effort into changes to systems, technology, and organization structure may not improve the quality, effectiveness, and efficiency of the work. Implementing E-Business requires wide-ranging, concentrated effort—effort that is technical, political, and managerial in nature. E-Business implementation requires elapsed time because the culture in the organization must be altered toward an E-Business culture.

E-BUSINESS ACTIVITIES

How Are E-Business Activities Different?

E-Business activities are integrated with each other. There should be a seamless interface that allows transactions to flow freely across boundaries. To a customer or supplier, E-Business may appear as one large business activity with many different steps. In E-Business there is less awareness of the business departments and units from a customer or supplier view. This is due to not only greater automation, but also the need to eliminate manual steps and exceptions in the work. E-Business requires more central coordination across departments; E-Business activities are distributed across departments. Without E-Business the IT (information technology) group plays a support role to departments; in E-Business IT plays a more central role. Systems and technology in E-Business are more integrated. You are more likely to outsource some of the implementation effort in E-Business due to the business pressure to implement E-Business fast. There is greater risk to the business since in E-Business you are exposed to customers and suppliers. Any problems and shortcomings are more visible to the outside world.

Figure 1.1 is a comparison table that will be useful in later chapters.

In addition to differences in implementation there are major ongoing differences. In E-Business you will be generating a stream of promotions, discounts, and marketing efforts. These will have significant impacts on how you perform and support the business. Management and marketing will demand flexibility to adjust to the marketplace. This will be new to IT and the organization.

Attribute	Standard Business	E-Business
Management view	Transactions support the business	Transactions are crucial to the business future
Management control	Distributed among departments	Centrally coordinated
Focus of implementation of change	Single business activities	Group of related activities focusing on suppliers or customers
Technology and systems	Separate, but interfacing	More integrated
Transaction exceptions, workarounds, shadow systems	Usually present	Must be reduced or, if possible, eliminated
Role of IT	Supportive	Larger role as an integrator
Outsourcing	Potentially used	More likely to be used
Potential risk	Limited since business activities are internally focused	Greater risk since there is more exposure to customers and suppliers

Figure 1.1   Differences between E-Business and Standard Business

E-BUSINESS ACTIVITIES ACROSS BUSINESS UNITS

Looking at the organizations within a company, you discover that many activities cross over multiple internal organizations. This may be by design and division of control and is often for a sound business reason. It simply may have evolved this way. An organization may have changed, with no thought given to the actual work. While you want to plunge ahead with E-Business, you do so in the context of the current business that is providing the money and support for your E-Business efforts.

In E-Business you have to consider groups of business activities that span the organization. You cannot just center your attention on a few activities. This could doom your E-Business initiative.

BUSINESS ACTIVITIES AND TECHNOLOGY

Let's explore the relationship between business activities and computer and communications technology and systems. The Internet, telephone systems, computer networks, electronic mail, voice mail, and facsimile machines are examples of infrastructure that support business operations. If the business tailors a computer software system to specific transactions, the computer system becomes integrated with them. The hardware, network, and other software necessary to run the systems are part of the infrastructure. In E-Business you cannot ignore such factors as facilities (e.g., offices and warehouses), location, and office layout. Unsuitable infrastructure can negatively impact E-Business.

E-Business typically requires new technologies and systems. While some of these may replace your current systems, often the new systems must interface to the current systems. In some cases, you can modify your current systems to support E-Business. However, this may be rare because your current systems may be older, legacy systems that were not designed to accommodate the degree of integration and response time demanded by E-Business.

FORMAL VERSUS INFORMAL BUSINESS ACTIVITIES

Formal transactions tend to be documented or at least acknowledged by the organization. Informal transactions and work, which are more numerous, may be either manual or computer-based. For example, people invent forms or put data into a PC. These informal "shadow systems" can lead to trouble. The organization becomes dependent on these systems, which usually are set up by a computer-savvy employee, and builds workflow around these. If the inventive employee leaves the organization, the informal transactions may fail or falter.

E-CRITICAL BUSINESS ACTIVITIES

In this book, you will consider activities that are critical to E-Business. An organization may view some (such as payroll) as more important than others. A business activity or function is "e-critical" if its performance significantly impacts revenue or costs of the organization for either E-Business or traditional business.

CHANGES TO A BUSINESS ACTIVITY OVER TIME

How does a business activity evolve over time? In many businesses, the business activity begins informally with manual transactions. The company grows and changes. No one pays much attention to the activity, because it appears to function. Over time, the volume of work increases, the nature and range of transactions change, and the amount of work per transaction increases. The people who originally performed the work and set up the procedures leave or disappear. People coming into the organization to do the work often do not receive formal training. There is no on-the-job training.

When you implement E-Business, some of the new procedures may have to be informal due to uncertainty and lack of knowledge. E-Business is not total automation although many people would like this. You will formalize these procedures later as E-Business takes off. E-Business transactions will change over time due to competitive pressure, technology change, and the experience of the business in E-Business. Automation makes transactions more formal and rigid as well as more efficient.

When people have implemented systems, the goal was often to support a part of a transaction or only a few key transactions that were high volume. Automation performed part of the work. When you implement E-Business, you will have to cobble together internal existing and new technologies to meet your deadlines. IT people have often concentrated on automating the easiest part of the business work. People then institute workarounds for the system to complete the processing of the transaction. The system itself begins to deteriorate into a maintenance mode. People discover that this cannot handle all of the requirements. There are additional management reporting and functional requirements. People then generate shadow systems. This is a fairly typical picture of how the work changes. In E-Business you must take a more comprehensive view.

Here are some signs of deterioration to watch for:

- The business activity may work, but it is unresponsive. Your E-Business grows and the technology cannot respond to the volume.
- There are no up-to-date, formal procedures. E-Business requires more formal procedures and policies to be automated.
- Typically, one or two people have been with the business activity for a long time. They can provide history but are ignored by management. This is true in both traditional and E-Business. These people are sometimes labeled "queen bees" or "king bees."
- A number of add-on, replacement activities may come to surround the basic transactions. This happens in organizations that attempt to respond to new marketing pressures. The result is a crazy quilt of patched procedures and policies.
- People may complain about how they do their work, but they fear change. The current methods are known and people are comfortable with them.

ACTIONS IN E-BUSINESS IMPLEMENTATION

Here is the sequence of actions in this book for you to implement E-Business:

- **Understand the business and select the activities for E-Business.**
  - — Action 1: Understand business goals and direction. Without this your E-Business efforts might not be aligned with the business.
  - — Action 2: Choose a set of activities for E-Business implementation.
- **Collect information—externally and internally.**
  - — Action 3: Assess the competition and industry. You don't want to be left in the dust by competition in E-Business.
  - — Action 4: Evaluate technology and infrastructure. Technology supporting E-Business is changing and new technologies are emerging.
  - — Action 5: Collect information for transactions, infrastructure, and organizations. Unless you understand what you have, you cannot implement E-Business successfully.
- **Define the new E-Business transactions and workflow.**
  - — Action 6: Analyze the current transactions and organizations. How does the current situation fit E-Business?
  - — Action 7: Define the new procedures and workflow for both E-Business and traditional business.
- **Get ready for E-Business implementation.**
  - — Action 8: Define and measure E-Business success.
  - — Action 9: Develop the E-Business implementation strategy.
  - — Action 10: Market E-Business to both internal and external audiences.

- **Implement the new transactions and workflow for standard business and E-Business.**
  - — Action 11: Plan the implementation.
  - — Action 12: Implement E-Business and modify the current procedures and workflow.
  - — Action 13: Follow up after E-Business implementation.

Implementing E-Business costs money and can generate controversy within the organization. Therefore, the E-Business implementation project must be well organized and measurable. Since E-Business implementation projects often cross multiple departments and involve significant organization and infrastructure changes, an implementation strategy is essential.

Also, analysis does not translate into results. Therefore, we have included actions for implementation planning, implementation, and postimplementation work. The detailed implementation plan supports the strategy. Postimplementation work supports measurement and often identifies where the effort should be directed next.

METHODS AND TOOLS FOR E-BUSINESS IMPLEMENTATION

A *method* is a technique for doing a specific piece of work on the project. Examples include collecting data, developing the new workflow and procedures, and testing these. A *tool* supports a method by making it easier and more effective to follow the method. Tools include transaction mapping, flowchart software, programming, testing, and simulation tools. Methods do not require tools, but are enhanced by them. Tools, implemented without methods, often fail, because people do not know the proper use of the tool. Tools in implementing E-Business have to be supported by training, experts, guidelines, lessons learned, and management expectations on what the tools are to achieve.

Implementation requires both methods and tools. Choosing the right ones is important. The wrong method or tool can be fatal to the E-Business implementation effort.

THE BENEFITS OF E-BUSINESS SUCCESS

Successful companies and managers cite a number of major benefits when E-Business projects are successful. Some of the more common ones are the following:

- *Reduced staff involvement in routine tasks.* With E-Business, customers and suppliers are performing some of the simpler tasks related to ordering, status checking, and tracking. Credit checking is reduced through automation. This frees up employees to concentrate on customer service and providing more advanced services or a wider range of products. In one firm, the company was finally able to gather business lessons learned on their products—something that they had wanted to do, but had no staff to do it. In E-Business you can redirect internal employees to more productive work.

- *Streamlined, simpler business activities.* When you implement E-Business, you really make the effort to drive out the exceptions and manual steps in transactions. You want to eliminate shadow systems.
- *Improved employee morale.* With many routine steps eliminated and steps gone, many employees often find that the new work is more creative and challenging.
- *Improved work quality.* With E-Business, since more is automated, there should be a smaller number of errors. The exception occurs when there are programming errors that undermine the E-Business transactions. This is why testing is so important in E-Business implementation.
- *More information.* E-Business brings a flood of automated, structured information on customer and supplier behavior and purchasing. Of course, there must now be new methods and procedures put in place to analyze the data.
- *Better decisions.* You tend to make better decisions with more information.
- *More flexible business policies.* Since you can implement new policies such as promotions faster, you gain greater flexibility in positioning your products. To take advantage of this, management must be more "fleet-footed."
- *Improved customer satisfaction.* If you combine the convenience of electronics with excellent customer service and competitive pricing, customers will be happier. They will also be more satisfied if you provide more information on the products and services to them.
- *Ability to enter new markets and reach new customers.* One firm we assisted was able to expand its reach from several countries to entire continents through E-Business. Another was able to offer new lines of products.

With all of these benefits you may wonder why E-Business implementation fails. Here are some common reasons for "e-failure."

- *High expectations.* Management gets lofty expectations based on the hype generated by the media for e-commerce.
- *Underestimation of effort.* Management gets the wrong perception that they can just implement the e-commerce software and then they will be in the land of E-Business.
- *Overreliance on methods that are laden with jargon and buzzwords and short on results.* The firm selects some vendor who offers some exotic techniques that may not be suited to the culture of the firm or the country. Many

efforts of these "e-engineering" firms result in e-failure. If you learn one thing in E-Business, it is that "culture counts."

- *The implementation project focuses on only a few activities.* Other supportive work is ignored—to the company's, customers', and suppliers' regrets later.
- *Overdependence on technology.* Some people think that if they buy some standard e-commerce software and hardware along with the network components, they are through. In fact, they are just starting.
- *Lack of consideration of the political environment.* Some firms that fail did not consider the internal political structure of the firm. They thought they could just implement the technology directly. If you do not gain major support, then you are more likely to fail in E-Business.
- *Over-control by vendor or consultant.* The organization turns the project over to an outside firm. The new approach never takes hold after the consultants leave. Some consultants want it that way so that they can stay forever and run the bill up.
- *Excessive upper management involvement.* This is the extreme in which managers attempt to micromanage the E-Business project.

E-BUSINESS OBSTACLES

Surveys have cited a number of barriers to successful E-Business implementation. Among the most commonly mentioned are:

- Coping with legacy systems and old technology
- Changing the internal organization culture
- Building an implementation team for E-Business
- Resolving conflicts between channels
- Upper management support
- Educating managers about E-Business and overcoming resistance

These will be addressed as we cover E-Business implementation.

DEBUNKING SOME E-MYTHS

Let us consider some common myths.

MYTH 1: CHANGE IS CONTINUOUS

It is true that E-Business implementation should be continuous. This would mean constantly striving to improve business activities in major ways. Unfortunately, this approach flies in the face of reality. People desire stability in their work. Observe that after a major change, the transactions tend to be stable. They then begin to evolve on their own. Just like the old procedures, the new ones need maintenance. They then begin to deteriorate. Thus, in E-Business you have to implement change in waves.

MYTH 2: FOCUS ONLY ON THE CRITICAL BUSINESS ACTIVITIES

The effort should focus on sets of activities containing some of the top "e-critical" work relevant to E-Business. Thus, even though you are working with e-critical activities, you have to take the time to include mundane and supportive work.

MYTH 3: TOP MANAGEMENT HAS TO PARTICIPATE ACTIVELY IN E-BUSINESS PROJECTS FOR THEM TO SUCCEED

Participation by management may be desirable, but is not mandatory. What is essential is management support during the entire project. You want to avoid excessive management involvement in detail. This can bog down your E-Business effort.

MYTH 4: THE ENTIRE BUSINESS ACTIVITY IS RADICALLY REDESIGNED

Radical change may not always be the best change—even for E-Business. In radical redesign of a process, the organization or the infrastructure (or both) may foil you. Radical change has even driven firms into bankruptcy. In E-Business some business activities may be enhanced. Examples are customer service and merchandise returns.

MYTH 5: THE PROJECT IS STARTED WITH A CLEAN SLATE

This might be good theoretically, but it ignores reality. Much of the organization and infrastructure must be accepted and taken as a constraint—even in E-Business. Who has the time and money to keep a business going and, at the same time, totally reinvent its way of doing business?

MYTH 6: THE GOLDEN METHOD OR TOOL WILL ENSURE SUCCESS

This is the vendor dream: their method or tool will solve all E-Business problems and result in new, widely acclaimed approaches to business. If the method is so great, why is not everyone using it? Methods and tools come and go. Do not pin all your hopes on a particular one.

MYTH 7: IF YOU DO THE ACTIVITY REDESIGN RIGHT, THE IMPLEMENTATION WILL BE SUCCESSFUL

What you learn in E-Business implementation may undo some of the original design for the business activities. Also, politics appear in implementation to a greater extent than in redesign, so changes may be required.

NINE PRINCIPLES FOR E-BUSINESS IMPLEMENTATION

A series of principles for E-Business implementation will be defined here and expanded on and supported in later chapters.

PRINCIPLE 1: A FIRM DEPENDS ON ITS BUSINESS ACTIVITIES FOR PROFITABILITY

To make money you must make something or provide a service. Doing this requires structured transactions and workflow. Companies that are successful in E-Business have integrated their organization and infrastructure with their work. The principle seems obvious. However, firms get carried away by thinking about their organization as opposed to how they do their work.

PRINCIPLE 2: BUSINESS ACTIVITIES SHOULD BE CONSIDERED IN SETS OR GROUPS

There are few stand-alone, separate business activities, and even fewer of these are critical. A good example is the large-scale acquisition of stand-alone personal computers in the late 1980s. Much of their value was not realized until networks were established so that the technology and people could be linked to the processes.

Often, the most successful companies have found a new and better grouping of activities when they implement E-Business. They then focused on having the organization and infrastructure support this group.

PRINCIPLE 3: BUSINESS ACTIVITY PERFORMANCE AND CHARACTERISTICS AS WELL AS INTEGRATION ALLOW YOU TO DIFFERENTIATE YOURSELF FROM COMPETITION

In E-Business it is not enough to have efficient work. The business activities have to work together. They have to supply a complete range of transactions in a timely and error-free manner.

PRINCIPLE 4: IMPORTANT NEW TECHNOLOGIES AND INFRASTRUCTURE SUPPORT CRITICAL E-BUSINESS AND TRADITIONAL WORK

Many potential technologies are irrelevant to E-Business. Even when they are relevant, their value is not automatically achieved. Adding a new technology creates gaps and problems in interfacing with the rest of the organization and infrastructure.

Consider the totality of all of the technologies supporting the business. Watch for a technology's lack of applicability. Many new technologies do not work because they cannot easily interface with other parts of the infrastructure.

PRINCIPLE 5: E-BUSINESS MUST BE MAINTAINED OVER TIME

This is often ignored. People assume that either a business activity is self-maintaining or that the people involved will automatically keep the transactions up-to-date. If an organization fails to maintain its transactions and work, it often withers and dies. Amazon.com, even with its leadership position, does not take anything for granted.

PRINCIPLE 6: E-BUSINESS IMPLEMENTATION IS BEST CARRIED OUT IN DISCRETE STEPS AS OPPOSED TO CONTINUOUS IMPROVEMENT

E-Business implementation requires considerable time, energy, and focus. If changes are made continuously, how can you be focused on anything else? Instead, improve the way business is performed and allow it to stabilize, be maintained, and evolve. It can then be revisited later as needed. However, note that overall, E-Business implementation and change are continuous. You move onto new areas of the business to expand E-Business.

PRINCIPLE 7: EMPLOYEES SHOULD BE VALUED BASED ON THEIR CONTRIBUTION TO THE CRITICAL BUSINESS ACTIVITIES

The operative word here is "should." Unfortunately, many companies reward people based on their value to the organization, not the work. Activities do not evaluate staff—managers do. Being valuable to the organization is different from being valuable to the work. One can be valued in an organization through politics and personal relationships. A business activity is not as emotional. When an organization rewards people for their contribution to the work, the work itself often will improve. Other employees recognize such rewards because their contribution to the work was evident and the importance of the transactions was obvious. In E-Business people are still crucial.

PRINCIPLE 8: BUSINESS ACTIVITIES ARE EASILY AND NEGATIVELY IMPACTED BY ORGANIZATION AND INFRASTRUCTURE POLICIES

You must have flexibility to change policies. In E-Business retailing this may mean that you have to be more flexible in return policies to retain customers. Your inventory levels may have to be adjusted to ensure that you have no substantial backorder situations.

PRINCIPLE 9: CHANGES IN ORGANIZATION AND INFRASTRUCTURE THAT ARE NOT MADE TO IMPROVE BUSINESS WORK TEND TO BE TRANSIENT AND SHORT-LIVED

For example, some people think that changing managers or getting the e-commerce software will somehow transform the work. It usually does not happen, unless the new managers move quickly to improve the business activities.

COMPARATIVE FACTORS FOR E-BUSINESS IMPLEMENTATION

In later chapters you will be building comparison tables that show the results of analysis. These can be presented to management and will help with the implementation of the new activity. The tables are based on the factors in Figure 1.2.

Business
> Business objectives—General goals of the business
> Business issues—Regulatory, competitive, and industry pressures

Industry
> Competitive and similar firms—Firms in the same industry or those with similar types of work and transactions
> Competitor's or other firm's business activity—Selected activity for analysis
> Industry averages/statistics—Ratios, trends, and statistics for the industry segment
> Competition infrastructure—General components of the infrastructure of the competitor
> Competition organization—Features of the organization (empowered, distributed, centralized, outsourced, etc.)

Organization
> Organizations (old/new)—Features of the organization
> Detailed organization (old/new)—Organizational units

Infrastructure
> General
> Internal infrastructure—Old/new
> Candidate infrastructure

Technology
> Architecture—Old/new
> Internal technology—Old/new
> Candidate technologies
> Degree of automation—Old/new

General business activities
> Single activities—All
> Activity groups

Activity group
> Work strategy
> Activities
> Transaction steps—Old/new
> Characteristics (age, etc.)—Old/new

Measurement
> Benefits—Estimated/actual
> Costs—Estimated/actual
> Performance measure (response, time, effort, paper)—Estimated/actual

Figure 1.2   Comparative Factors for E-Business Implementation

HOW TO USE THIS BOOK

Each of the E-Business implementation actions is presented using the following structure:

- *Introduction*—background for the action.
- *Milestones*—end products of the action.

- *Methods for E-Business*—the details of what to do.
- *E-Business Examples*—what some firms have done in E-Business implementation.
- *E-Business Lessons Learned*—lessons learned from firms implementing E-Business.
- *What to Do Next*—things you can do right away.

To support the overall sequence of actions, you will build a series of tables that show relationships between the business, transactions and work, organization, infrastructure, competition and the industry, suppliers, and customers. The tables are based on the comparison factors defined in Figure 1.2. Constructing tables will help with understanding, analysis, decision making, and action.

E-BUSINESS EXAMPLES

We will draw upon four examples of companies who were successful in E-Business. While the names of these firms are hypothetical, they are based on combinations of real world firms. In implementation each faced a series of major decision points and crises. All were successful.

Ricker Catalogs was a standard producer and distributor of paper catalogs. They had a loyal customer base. In their catalogs, they sold their own products as well as those of other suppliers. At the start of E-Business implementation, there was no internal experience. Ricker was drawn to E-Business for both negative and positive reasons. Management saw E-Business as a chance to reach new customers and gain new suppliers. They also were aware that the catalog business as they knew it may well shrink due to E-Business. On the other hand, by embracing E-Business, they could become one of the darlings of the e-stock market hype. They chose the wrong strategy first and then had to change. You can learn a lot from this firm's experience.

Marathon Manufacturing is a major manufacturer of machines and components for metal fabricating products. Marathon had concentrated on reaching customer firms that had at least 50 employees. They decided to enter E-Business to reach the hundreds of thousands of smaller firms. They decided to take the time and do the web site right while changing their internal business transactions. This is a case of success from the beginning in which there was no time pressure.

Abacus Energy is a major international oil company. At first, they did not see the opportunity or need for E-Business. Their problem was a lack of vision as to the potential of E-Business. They were focusing on their customers to whom they sold gasoline, other petroleum products, and food products. Finally, they decided to consider their suppliers. The E-Business project here was the replacement and improvement of the transactions for purchasing and contracting. Although this has a narrow scope, it will be useful to us as an example.

Crawford Bank is a major domestic bank that considered using the Internet and web to offer installment loans. They hoped to build new business and reach new customers through E-Business. Using these customers, they hoped to build an entire new company. They decided to start small and then grow.

CRITICAL SUCCESS FACTORS FOR E-BUSINESS IMPLEMENTATION

There are certain critical E-Business success factors common to all four firms.

- Management was flexible in terms of changing policies as long as the current customer base was not impacted.
- Management did not excessively interfere with the E-Business implementation.
- While using consultants, they did not over rely on them.
- The traditional business activities were integrated or at least linked with E-Business.
- The firms involved a wide spectrum of employees in their E-Business implementations.

GLOSSARY

Here are some of the technical terms used in this book, along with a definition of how they are used:

- *Acceptance testing.* Testing performed by business staff of the new workflow, transactions, and systems.
- *Activity group.* Business activities that are related through one facet of a business, such as organization, technology, suppliers, or customers.
- *Benchmarking.* The development of comparisons of your firm to others.
- *Black box testing.* Testing in which you assume inputs into, and outputs from, a transaction.
- *Comparison tables.* Tables constructed in all steps to assist in analysis and marketing of the work performed for E-Business implementation.
- *Deterioration.* The decay of a business activity over time.
- *Enhancement.* Improvements to E-Business to meet new requirements—something that you will doing forever.
- *Exception.* A variation of the business workflow to address a specific transaction. Exceptions tend to negatively impact E-Business.
- *Facilities management.* This refers to the management of a firm's computer

systems by another firm. You may consider outsourcing much of your E-Business infrastructure.

- *Infrastructure.* The facilities, equipment, and other support for E-Business including warehousing, facilities, and layout.
- *Integration.* The interfacing and combining of the procedure, technology, and system components. Successful E-Business means successful integration.
- *Maintenance.* Changes to the work to meet the original requirements. All E-Business business activities require maintenance.
- *Method.* A set of procedures for performing some function. E-Business implementation relies on reliable methods.
- *Organization.* The business's structure and its employees—key ingredients to E-Business success.
- *Performance testing.* Testing of the E-Business systems and technology to ensure that response time and workload are within acceptable limits.
- *Pilot.* The testing and review of both the prototype system and E-Business.
- *Prototype.* The initial version of the system that will support the E-Business.
- *Scenario.* One possible version of the new business activity and its transactions.
- *Scorecard.* The criteria for evaluation of a business activity.
- *Shadow system.* A manual or automated system created to handle work outside of the systems supporting the business activity.
- *Strawman.* A model or sample of the new E-Business activity.
- *System testing.* The overall testing of the computer system that supports E-Business.
- *Technology.* All categories of technology involving computers and communications.
- *Tool.* A tool supports the performance of a method.
- *Transaction testing.* Detailed testing of transactions involved in the new procedures and system.
- *White box testing.* Testing of a component of a business activity or system internally.

WHAT TO DO NEXT

Complete the following questions or use them as the basis of a short survey of managers to determine which activities managers and staff view as e-critical. These questions also raise interest in E-Business implementation by getting people involved.

1. Identify 10 business activities that are e-critical to E-Business. Rank these in order of importance to the business (1 is highest; 10 is lowest).

1. _____   2. _____

3. _____   4. _____

5. _____   6. _____

7. _____   8. _____

9. _____   10. _____

2. Now take the same activities and rank them in terms of degree of problems and potential improvement on the same scale.

1. _____   2. _____

3. _____   4. _____

5. _____   6. _____

7. _____   8. _____

9. _____   10. _____

3. Compute a total score for each activity by multiplying the ratings for importance and potential for improvement. This approach helps you identify activities that rank highest overall.

Activity Score

1. _____   _____

2. _____   _____

3. _____   _____

4. _____   _____

5. _____   _____

6. _____   _____

7. _____   _____

8. _____   _____

9. _____   _____

10. _____   _____

4. Based on your knowledge of business procedures and workflow, indicate the degree to which you agree or disagree with the following statements on a scale of 1 to 5 (1, strongly agree; 2, agree; 3, indifferent; 4, disagree; 5, strongly disagree):

The most severe problems with the current business lie in

Organization	1 2 3 4 5
Information systems	1 2 3 4 5
Infrastructure (buildings, telephones, etc.)	1 2 3 4 5
Lack of management interest	1 2 3 4 5
Age of the business activity	1 2 3 4 5
Bad information in the business activity	1 2 3 4 5
Lack of formal procedures	1 2 3 4 5
Staff turnover	1 2 3 4 5
Lack of procedure training	1 2 3 4 5
Interfaces between departments	1 2 3 4 5
Interfaces with suppliers	1 2 3 4 5
Interfaces with customers	1 2 3 4 5
Integration with other systems	1 2 3 4 5

5. For companies that have not been successful in previous work improvement efforts, examine the following common reasons for problems in carrying out these efforts. Rate these on a scale of 1 to 5 (1 means it does not apply at all; 5 means it strongly applies). Your responses will help you determine what obstacles you will later overcome.

Lack of management involvement	1 2 3 4 5
Excessive management involvement	1 2 3 4 5
Method was oversold, raising excessive expectations	1 2 3 4 5
Lack of project management	1 2 3 4 5
Lack of measurement of the current activity and system	1 2 3 4 5
Project to implement was too large	1 2 3 4 5
Overdependence on consultants	1 2 3 4 5

Your E-Business Implementation Roadmap

E-BUSINESS CHALLENGES

There is intense pressure on many businesses from different directions related to E-Business. Some of these are:

- Pressure from stakeholders for improved financial performance through E-Business.
- Competitive pressures. Your firm may face threats from not only standard competitors, but also from new competitors who seek to enter your territory.
- Global regulatory changes. Changes are in the works continually by nations to make their companies more competitive in E-Business.
- Technology. You are faced with the problems of keeping up with changing information and communications technology.
- More demanding customers. Customers are placing new demands on firms for products and servicing.
- Cost pressure. Firms are being pressured to cut costs and improve productivity through E-Business.

In addition, the project manager in E-Business implementation faces challenges that are unique:

- Strategy. You must develop the E-Business strategy and then implement it.
- Continuous E-Business implementation. Unlike most projects, E-Business implementation does not end. The first E-Business implementation is just the beginning.

- Technology infrastructure. The technology must integrate and support the business activity as well as the implementation of E-Business. Typically, this means integrating E-Business technology with your existing technology.
- Management. Coping with management expectations and involvement involves significant challenges.
- Staff. First, there is fear among the staff involved and affected by E-Business. Second, there is fear among the team members in regards to their future after the project is completed.
- The new business activity and organization. Specifying, designing, and developing the workflow, transactions, and the organization approach for E-Business are a challenge.
- A steady stream of results. Expectations are high. They continue to be so (and even rise) if some success in E-Business is achieved.

FACTORS AFFECTING E-BUSINESS SUCCESS

Some of the most important factors that affect the success of the E-Business project are the following:

- Making the case to implement. This is the marketing and sales issue.
- Dealing with changing requirements during E-Business implementation from technology, management, and the competition and marketplace.
- Addressing the cultural issues of the organization that affect the E-Business project.
- Establishing and keeping the right level of management attention and involvement during E-Business implementation.
- Ensuring that the infrastructure supports E-Business.

Keep these in mind during implementation.

DEFINITION OF MAJOR DELIVERABLE ITEMS

A minimum list of deliverable items has been developed which sets apart serious efforts at E-Business implementation from those that produce stacks of reports. Here is a list of those items, along with the action from Chapter 1 in which you would generate the item:

- The recommended set of business activities to be improved to implement E-Business (Action 2).

- The assessment of external information (competition, technology) and internal business activities that relate to E-Business (Actions 3–6).
- The new E-Business activities, transactions, and workflow—how E-Business will work, the benefits, and what changes are required to make it come true (Action 7).
- The implementation strategy for E-Business (Action 9).
- The E-Business implementation plan (Action 11).
- Implementation of E-Business (Action 12).
- The results obtained from E-Business (Action 13).

If you expand this list, you risk being diverted into side trips that may not contribute to your E-Business success. If you attempt shortcuts, then you risk failure and redoing the work.

YOUR E-BUSINESS STRATEGY

You want to define an overall strategy for E-Business. Four general strategies you want to consider are:

- *Separation.* E-Business as a separate activity.
- *Overlay.* E-Business implemented on top of standard business.
- *Integration.* E-Business integrated with your regular business.
- *Replacement.* Replace some of the existing regular business transactions with E-Business.

E-BUSINESS STRATEGY ALTERNATIVE: E-BUSINESS AS A SEPARATE ENTITY

There are arguments to support each of the previous strategies. Establishing a separate entity can be attractive if you wish to address different customer segments than your regular business. You also might be able to get the E-Business established faster since you will not be encumbered by your existing business practices. However, this strategy can be more expensive. You might have to implement redundant business activities. This might include staffing, systems, technology, and infrastructure. Crawford Bank decided to offer their installment loans through the web as a separate entity so as not to drive current customers onto the web. A leading retailer started with the overlay strategy and then moved to the separation strategy.

E-BUSINESS STRATEGY ALTERNATIVE: E-BUSINESS ON TOP OF REGULAR BUSINESS

Many firms elect to implement E-Business on top of their standard business. This seems at the start to be a quick approach that gets your web site up quickly. The argument is that you can leverage off of your existing business. However, there are drawbacks to this approach. Normally, you will have to change the internal culture of the firm toward E-Business. This can be threatening to the organization and its employees. With this strategy you will also have to modify some of your e-critical activities to handle both normal business and E-Business. Many automobile firms are examples of this. Some banks have followed this strategy for their credit card operations. Ricker Catalogs started with this approach. It led to many problems. Consideration was then given to the separation strategy. This was rejected as being too expensive. Ricker then opted for the integration strategy.

E-BUSINESS STRATEGY ALTERNATIVE: E-BUSINESS INTEGRATED WITH NORMAL BUSINESS

This is the approach that Marathon Manufacturing followed. They decided to take the time and spend the resources to develop and implement an integrated strategy. It paid off. This is the most complex strategy for E-Business and it is addressed in this book. The integration strategy is also the most potentially rewarding. Complexity arises because you have to change your current business practices while at the same time implementing E-Business and keeping the existing business going—a real challenge.

E-BUSINESS STRATEGY ALTERNATIVE: E-BUSINESS REPLACES STANDARD BUSINESS

Here you would select a set of business activities for E-Business. You would then change these to an E-Business structure. The old activities are replaced. Only E-Business transactions are used. Abacus Energy chose this strategy for purchasing and contracting. Their goal was to support outsourcing as a business strategy. E-Business implementation resulted in substantial cost savings, improved supplier–Abacus relations with business departments, and support of more effective outsourcing.

E-BUSINESS PROJECT CONCEPT

Before you begin the planning for E-Business, you should define the project concept. The project concept includes the following:

- Business and technical purposes of the project
- Scope of the project
- Roles of the business units and IT in the project
- Issues that you are likely to face in the project
- General cost and schedule of the project
- Major milestones of the project

The project concept sets the stage for the project plan. It aims to gain consensus and a common view of the project. This is particularly necessary in E-Business projects where there may be many opinions.

The scope of an E-Business project can expand and contract with the progress of the work, requiring a flexible project plan. The project will begin with a low-level analysis of the business. Then a candidate group of activities will be defined. The project will expand to a more detailed external and internal data collection and analysis. This leads to the development of the new business activities. Implementation there can bring more changes and redirection.

The scope can also change due to changes in management policy, organization, and other factors (both internal and external). Organization and infrastructure changes may set off a political battle. The project may be forced to go underground if a significant organizational battle surfaces.

Issues can arise that must be addressed and may change the project's scope. Outcomes of issue resolution often impact the plan. You want to get out on the table at least 40–50 issues that relate to organization, transactions, policies, customers/suppliers, management, and technology.

E-BUSINESS PROJECT PLANNING

OVERALL PROJECT TASK PLAN AND SCHEDULE

This is a standard set of tasks and milestones that includes the following three areas: dependencies, resource assignments, and duration of tasks.

- **Dependencies**
  Two tasks are dependent on each other if one must be started or completed before the other can begin. Give close attention to the structure of the tasks

and dependencies. Assign dependencies between tasks at the same level, which will make it easier to change the schedule.

- **Resources**
  Figure 2.1 gives a candidate list of resources. Assign only a few key resources per task. Assign resources at as high a level in a task outline as possible to reduce maintenance and detail, and never identify resources by exact name, but instead by role or function.
- **Schedule**
  The schedule is a valuable tool for addressing issues and politics. Each major area in Figure 2.2 can have a separate schedule. Experience shows that during an E-Business project, additional resources will be offered, resources will be removed, and people will be reluctant to change organization and infrastructure. To get these changes through and to maintain momentum, show how an issue impacts the schedule. GANTT charts are appropriate here.

Update the schedule and plan regularly. This increases credibility and establishes management's trust. If you simply build the project plan for project approval and never show it after that, you risk failure. The schedule should contain the original baseline or established schedule. Changes and updates produce the current schedule. Be able to explain the reasons for the changes.

Here are some suggestions on generating an E-Business implementation project task plan:

- Ensure that members of the project team have the opportunity to define the tasks. Participation generates enthusiasm. Identify task areas and leave the detail in each area to a team member.
- Flesh out the project task plan for the current phase of work and the next phase ahead.
- Work hard to build an adequate level of detail to prevent issues from arising. Issues are more likely to occur in low-level tasks so keep a close watch on those.

- **Tools**
  — Project level
  — Business activity level
  — Organization level
- **Organization**
  — Full-time roles
  — Part-time roles
  — Roles of contractors and consultants

- **Management**
  — Resolve issues in E-Business implementation
  — Review milestones
- **Project team**
  — Roles
  — Skills needed

Figure 2.1   Sample List of Resources

- Apply resources to tasks. This is important, since you may be required to defend your use of resources. Also, show how slippage in infrastructure and organization change can impact change and the overall schedule.
- With politics involved, tasks must be clear and nonpolitical. After all, the task plan is a document that can be employed for political self-interest.

People usually think of a team as a group of people who will be on the project throughout its life. In E-Business projects this is *not* true. Instead, there are three levels of team players:

- People who are formally assigned to the project full-time. There should be very few of these until full implementation. Too many full-time members can lead to inertia and inflexibility.
- People who are informally available to work on the project with the proper political negotiation. This category includes the senior people involved in the current work (e.g., queen and king bees).

Definition of business activity strategy
- — Activity selection
- — E-Business architecture
- — E-Business priorities
- — E-Business implementation
- — Measurement strategy

Implementation plan
- — Business activity
- — Organization
- — Infrastructure
- — Conversion and training
- — Testing
- — Completion of plan—(milestone)

Implementation of—infrastructure
- — Developed by area specific to project
- — Divided into facilities, equipment, technology, and support

Implementation of—organization
- — Developed by organization; likely to be general due to politics

Implementation of—business activity
- — Business activity changes
- — Business activity change support (—procedures and, standards)

Post-implementation activities
- — Assessment of benefits and costs
- — Determination of next areas for E-Business
- — Definition of lessons learned

Continues

Figure 2.2   Sample List of Tasks for an E-Business Project

Project management tasks
— Determination of the level of plans
— Evaluation and selection of methods and tools in project
— Implementation of methods and tools
— Establishment of project team
— Determination of review method
— Issue management approach
— Project reporting method

Identification of e-critical business activities
— Assessment of general business
— Development of list of e-critical activities—(milestone)

Selection of business activity group for E-Business
— Identification of alternate groups
— Addition of subsidiary work and transactions
— Definition of full groups
— Evaluation criteria for activity groups
— Evaluation of groups
— Selection of the activity group—(milestone)

Assessment of the activity group
— Industry and competitive assessment
— Technology and infrastructure assessment
— Evaluation of current business activities—(milestone)

Development of new E-Business transactions and workflow
— Development of alternates
— Definition of perspectives for evaluation
— Evaluation and selection
— Development of scenario of the new workflow and transactions—(milestone)
— Assessment of benefits and differences—(milestone)

The prototype and pilot of E-Business
— Prototype structure
— Initial implementation of the prototype
— Monitoring of the prototype
— Lessons learned/activity modification—(milestone)
— Selection of pilot project area
— Pilot project implementation—of infrastructure
— Pilot project implementation of—business activity
— Evaluation of pilot project/lessons learned—(milestone)

Determination of measurement methods of business activities
— Definition of measurement criteria
— Measurement data collection and analysis procedures
— Measurement methods and tools
— Measurement of existing business procedures and workflow
— Measurement of E-Business
— Cross-activity measurement

Figure 2.2 Continued

- People who perform specific tasks in the project and then leave. Given the scope of E-Business implementation, it is valuable to have a number of people with a wide range of skills.

It is important to keep the core E-Business team small. In Ricker's case the core team was three people. For Marathon it was five people; Crawford had three; and Abacus had four. Larger teams carry more risk and visibility and are more difficult to manage. More coordination is necessary.

Initially involve people from outside the immediate organization or externally to provide a fresh view. Using outsiders is valuable as long as they do not want to become players in the business activity or organization. Beware of people from outside the organization who ask to be on your project. They may have a hidden agenda.

For the core team, look for people who are generalists and can deal with the project environment. Team members have to be able to work together. Take people on the team for a trial period to see how they fit with the rest of the team.

Consider a mixed strategy of diffusion and concentration of team effort. At times there is a great deal of work that calls for parallel, diffused effort. When a major issue or hurdle arises that the team can address, get focused. No group of people can stay constantly up and in a team mode. Rather, team members should work as individuals most of the time. The project leader should consciously control this strategy as much as possible.

People want to know their duties and responsibilities. Be as detailed as possible with respect to their roles. Identify minimum rather than most desired effort. Negotiate with line management to delineate exactly what each person will do.

Be aware that you may not be able to get the best employees full-time. People can get burned out during these projects, so anticipate replacement. Rather than get the best people, get more average people who have energy. They will be less likely to be removed from the team.

THE E-BUSINESS PROJECT LEADERS

Generally, you will have more than one project leader given the scope of E-Business implementation. This is due to the range and extent of work for E-Business implementation. An E-Business project leader should have the following characteristics:

- Takes a broad view and favors change as opposed to minor enhancement.
- Tackles issues and opportunities during projects.
- Has been involved at the detailed level of a business activity or technology.
- Feels comfortable with informal contacts with upper management.
- Has performed a variety of duties which not only provide an individual with

an overall view of the company, but also ensures a wide perspective across divisions.

- Is known in a positive light to people in different divisions and departments.

The project leader should actively do work and not just administrate. On one E-Business project the first recommendation was to replace the project leader with a doer, as opposed to an administrator.

MANAGEMENT

The ongoing support of upper management is so important that management communications and presentations are frequently addressed. Resistance (both overt and covert), inertia, and the need to keep the business operating all combine against E-Business every step of the way. Without continuing management support, the project often falters.

On the other hand, managers have their normal jobs. Be aware that management's role may be limited to lending crucial support in specific crisis situations. They do not have the time or energy to be heavily involved in the day-to-day detail.

METHODS AND TOOLS FOR THE E-BUSINESS PROJECT

A method is an approach to guide a specific project activity. Tools support methods. Using project management software is a method. The actual software is a tool. Determine methods for the following activities:

- Identifying and managing issues
- Tracking the project
- Dealing with crises
- Managing and communicating with the team
- Collecting data and performing analysis
- Reporting to management
- Documenting the old and new transactions and workflow
- Simulating alternative actions and estimating results
- Assessing the infrastructure and organization
- Approaching the implementation of E-Business
- Measuring the effectiveness of the method

The project leader should directly address each of these. If the leader and team improvise methods, not only is it wasteful, but it also risks the team's credibility.

For each tool you are considering, answer the following questions:

- In what circumstances is the tool appropriate? Tool use depends upon the situation and the time required for the effective use of the tool.
- How will people learn about the tool and gain expertise (not just knowledge)? Sending people to training classes is not sufficient; they require hands-on use soon after training.
- Who will be the resident expert on a specific tool? The expert will ensure that people use the tool correctly.
- Under what conditions will a tool be replaced or cease being used?
- How will the tool's effectiveness be measured?
- How will the tool be described to others in nontechnical terms?

For most E-Business efforts, use a standard project management method. Select software that is a popular, endorsed product. If you pick something unfamiliar, you will spend hours defending your choice. Use a simple tool that costs a few hundred dollars at most. Evaluate a tool by how fast you can gain proficiency, not by the range of features. Get in the habit of using the software yourself to generate status reports and do "what if . . . ?" analyses.

How you organize the information you collect is of prime importance. Failure to be organized at the start will mean problems later. Maintain a log of activities and meetings. Retain all project correspondence. For tracking issues, set up an issues database that will track each issue, along with its status and resolution. When an issue recurs, you will be ready.

For documenting the old and new business activities, use simple graphics software rather than an exotic tool. Remember that the underlying purpose for the documentation is to increase understanding and support.

MANAGEMENT OF E-BUSINESS ISSUES

You need a method and a tool for managing issues. In E-Business projects, it is important to not only solve issues, but also to market and sell the solution. Difficult or complex solutions do not sell themselves. Also, if issues are addressed in haste they may quickly recur.

The method for tracking and managing issues could be software. The tool could be a database application on a PC. An issues database can be linked to specific tasks in a project plan under project management software to enable you to assess the impact of issues.

MANAGING AND REVIEWING THE E-BUSINESS PROJECT

Beyond the normal project tracking of budget and resource consumption, make sure that you are aware of the state of the project and progress being made. Be aware of what every person on the team is doing. Get informal status directly from people. Seek outside perceptions as to how people think the project is doing.

COMMUNICATING DURING THE E-BUSINESS PROJECT

A project leader faces a number of challenges in communicating. One challenge is the need to communicate the scope of the project and its differences from other projects. The project must avoid being typecast.

Communications must be frequent. Keep someone informed on a twice-a-week basis as to what is going on with the work. Even if work is normal and there are no major issues, take time to inform management of what is happening in the project. Informal status and progress reports build trust and relationships. An issue can be presented but identified as one not requiring immediate action. Later communications and meetings can address the issue after people have had an opportunity to give it some thought. This avoids surprise. There is always a tendency to use jargon. E-Business has its own—mostly beginning with "e." Avoid it. Keep to standard terms.

Rumors will abound. Make sure everyone is aware that you want to hear them. Become a one-person rumor control center. Disseminate information that addresses the rumor indirectly. Never give the rumor credibility by bringing it out in the open. To do so places you in a reactive mode.

The message you portray on the project must be consistent and relate to the business activities. It should not appear to fluctuate between organization and infrastructure. These elements support the approach; they are not the approach itself.

On a large E-Business project, it is difficult to maintain enthusiasm, but you must try. Trumpet small successes in thinking and approach. Give credit to people involved in the implementation. Greater involvement gives rise to enthusiasm.

For project communications, one-on-one meetings are recommended for anything significant related to issues. With the issues defined, divide them into two areas: those that are politically sensitive and those that can be addressed openly. Deal with political issues in a quiet manner. It is useful to have other issues surface. This shows progress and activity and gives people an opportunity to own the change method. For communications on nonpolitical, less important items, consider voice mail and electronic mail. Do not overuse these mediums, and keep the message short. Make notes, or write out the entire message first. Write to a draft file and then review it. Casual messages cause more harm than good.

REVIEWING AN E-BUSINESS PROJECT

In the early stages (up to the selection of the activity group), the project is inexpensive and should be low profile. Have the project reviewed by a manager.

The role of the project leader at this point is to provide status to managers on a one-to-one basis.

After the activity group is selected for detailed analysis, people will be more aware of it. This is a good time for a committee of managers and staff from the affected activity group to become involved. This gives them an opportunity for ownership and a sense of participation. Keep the committee informal, as a formal committee may burden the project with overhead.

During the implementation of the changes, use two review groups. One could be derived from the committee formed to deal with implementation. A higher-level group could deal with strategy and general E-Business issues. What should you do as a reviewer? Cast an eye on the scope of the E-Business project. What has changed? Look at the project factors before considering standard items such as budgeted versus actual allocations. Progress is not always evidenced by traditional work, but may involve solving problems related to policies, organization, or political conflict.

Some have suggested that a steering committee, consisting of a group of managers, should meet to review the progress of the project and resolve issues. This approach, however, has several pitfalls:

- A steering committee makes the project visible. In early stages, high-level visibility is useful, but during implementation, lower- and mid-level involvement is better. Greater visibility means higher risk and exposure.
- The review requirements of the project vary with the stage of the project. The same steering committee should not work on every stage.
- A steering committee assignment is political. If higher-level managers see that their peers are sending lower-level personnel to meetings in their place, they will also send substitutes. The result can be a lowest common denominator of employee.
- A steering committee makes the review method too formal. Most often, the project benefits from informality.
- Timely decisions may be difficult to achieve. Some E-Business projects may require only a few critical decisions. Get the support of the right manager. Give that manager the opportunity to sell the decision.

MEASURING EFFECTIVENESS

Evaluate the effectiveness of project management by various methods. Ask the following questions:

- How do the requirements and design of E-Business activities match up to the competition?
- Is the team working together?
- Where are time and resources being expended?

- How many and what types of issues remain unresolved over time?
- Is the schedule realistic?
- To what extent is the project team relying on methods rather than on tools?

GETTING MANAGEMENT APPROVAL

Key presentations relating to the approval of the project and the implementation project plan are explored in detail in the marketing chapter. At the start of the project (even when you are developing the plan), people expect you to have a great deal of detail. Be ready to respond to detailed questions at the front end of the project. Be purposely vague about the tasks and work that are further out in the future. After all, you do not know which business activities are to be improved.

Be able to answer questions about project costs and benefits at the same time. However, make every effort to separate these from the schedule and the plan. The plan is the "how" you will do it. The costs and benefits are the effects of carrying out the E-Business project. For example, if someone says that some part of the project is too expensive, focus on the work itself and determine if there is any better way to do it. Always remember that costs and benefits are the impacts of the schedule.

Always track cumulative costs and benefits. Infrastructure change can absorb the largest part of the budget, so make sure that all infrastructure improvements and work map directly into benefits for the business activity. Infrastructure costs tend to occur at the front end. Benefits occur later. If you create a curve of benefits minus costs, the curve will be negative for some time. It is important to track when this curve goes positive. The negative value should give you a boost to obtain management support for organization change to take advantage of E-Business efforts.

Within the project itself you will be tracking the actual versus the planned budget. This helps prevent diversions or delays in the project. During many of these projects, interesting opportunities surface as you look at a business activity in detail. Have the actual versus planned budgets ready so that you can keep on track.

Here are some tests to apply to your work. Did you develop a project plan that was even and consistent in level of detail? Did you write a series of issues that will have to be faced in the E-Business project?

To measure both the plan and your list of issues, find all tasks in the plan that correspond to the issues. Label these as having high risk. Next, go through the tasks and try to find other tasks that are risky. If you find a task that has risk and no issue, create another issue. If you have an issue without associated tasks, either the issue is not important or you are missing some tasks.

Confronting major issues with management is difficult. Here are some ways to deal with a crisis:

- Continue the project as is. In the midst of major upheaval, this alternative is infeasible. Much of the completed work will have to be redone.
- Slow down the project, but maintain momentum. This alternative is not attractive, because resources are still being consumed and morale is plummeting.
- Accelerate the project to force resolution and change. Accelerating the project is a practical approach. In some cases, issues involving business activity and organization are vague. More attention, pressure, and effort may force the issue into the open. This saves money in the long run and prevents the E-Business project from dissolving into minor efforts and web pages.
- Halt the project and examine the issues. This forces people to address the issues because they will have to be reassigned if the issues are not resolved. This action takes management initiative and incurs risk. If there is no project change and work resumes, the project is more likely to fail.
- Terminate the project. Terminating a project is a major decision. It is not the ending of the project that is a consideration as much as what will follow. Some companies immediately start another E-Business project after failure. This shows management's intent to get results. Many companies lack the will to do this. Management quietly drops the project and goes on about their business. Employees may think that management is not serious about E-Business. Staff members who volunteered and participated in the project will be reluctant to get involved in another project and another failure.

When you terminate any project, it is important to move quickly to identify lessons learned. This does not refer to placing blame but rather means examining the entire range of the project to learn how to prevent future failure.

E-BUSINESS EXAMPLES

RICKER CATALOGS

Ricker management thought about implementing E-Business for some time. They initially thought that they did not have the money to pursue either a separate organization or an integrated approach. They selected the overlay strategy where E-Business is layered on top of the regular business. They thought that they could implement E-Business faster. There was pressure from investors to get into E-Business. Ricker feared that if they moved too slowly they would be left behind.

MARATHON MANUFACTURING

No one in Marathon's industry segment had considered E-Business. Their segment was very traditional. Marathon decided to investigate the potential for E-Business. They appointed two project leaders—one from the business and one from IT. After some study Marathon decided to implement E-Business by changing the current business activities and implementing E-Business. They also consciously decided to involve as many employees as possible in the effort.

ABACUS ENERGY

Abacus Energy did not see the potential for E-Business for customers. However, they were finding that their supplier relations were quite poor. They decided to focus on suppliers in their E-Business efforts. Their strategy was to change the current purchasing and contracting practices, transactions, and workflow into E-Business.

CRAWFORD BANK

Crawford Bank was a successful bank. They saw clouds on the horizon in the form of new banks that offered discount loans on the web. Rather than start a new separate business, they decided to establish a new channel through the web. The long-term strategy was to establish a separate firm.

E-BUSINESS LESSONS LEARNED

- **Adequate and credible motivations are necessary to carry out change.** Change for the sake of change is not a reason for E-Business implementation. Tearing everything up and starting over is too disruptive to take lightly. People must understand the importance of E-Business. They must be motivated to champion change and E-Business.
- **Emphasize that E-Business implementation will occur in waves.** Continuous change and evolution is too destabilizing for most organizations to tolerate. As changes are implemented, revisit them for further improvement.
- **Base progress reports on results and recommendations, not on level of effort and activity.**

E-Business implementation projects are never level-of-effort standard projects in which a specific set of resources is committed. They are based on substantial breakthroughs in thinking. A typical E-Business project offers many small victories for management to enjoy and savor. The review method should allow for this time to "smell the roses."

- **If people cannot understand a method or tool in simple English, avoid it.**
  There is a direct relation between the use of arcane jargon and symbols and time spent in explanation and training. This time should be spent in marketing or getting agreement. Training is overhead, and in an E-Business project, with tight time and resource schedules, it is very expensive overhead.
- **Have a strategy ready for the manager who runs to you with the latest buzzword from a seminar.**
  Get more details, analyze what the manager says, and return with an assessment of how this would or would not fit with the project. This strategy conveys that you value what the manager says.

WHAT TO DO NEXT

1. Develop a general list of 50 tasks for an E-Business implementation project using the materials in this section. For each task enter the resources and predecessor task, if any. Enter these into a project management system through a copy and paste method.

2. For a sample business activity with which you are familiar develop a list of people who could serve as core members of the E-Business project team. Identify their roles as well as what they are working on now. Think about how you could get them on your team.

Person Project Role Current Work

1. _____
2. _____
3. _____
4. _____
5. _____

3. Now take a wider view and think about people you will need part-time for specific tasks based on their technical expertise or business knowledge.

Person Expertise Time Needed

1. _____

2. _____

3. _____

4. _____

5. _____

4. Listed below are some common objections encountered in recruiting people to join the project. Think about which apply to the people you have listed. Also, begin to think of arguments that will overcome these objections.

- Work on other projects has priority.
- There appears to be nothing in it for me.
- I have never done an E-Business project before.
- It is likely to fail so why should I get involved?

5. Given your knowledge of managers, identify who might serve as a project sponsor and who might serve on a steering committee. Also, indicate how much of their time would be needed.

Sponsor: Position: _____

Time needed: _____

Steering committee Time needed: _____

Name: _____   Position: _____

_____

_____

_____

6. What management style would you use for updating the committee and presenting issues for decisions?

7. What is the latest management fad or concept that your firm attempted to implement? What was the result? Were there tangible benefits? How were results measured?

8. What project management software tool do you have available? How proficient are you in the use of the tool? Is the tool endorsed by management? Are other tools in use? What political risk is involved in using the tool?

9. How are project issues resolved by management? What approach would you use for an E-Business issue?

Part II

Collect Information on the Business Processes and Technology

Chapter 3

Action 1: Understand Your Business

INTRODUCTION

Even if you are going to implement E-Business as a separate entity, you must select the business activities that you will implement. You are seeking a logical grouping of business activities for E-Business. You will not only select activities, but you will also support your choice against a backdrop of all business activities and their importance to E-Business and the business in general. To do this, you first need to understand the business. In this first action you would like to answer the following questions:

- What are characteristics and trends in the industry in general related to E-Business?
- What are trends in the implementation and results using E-Business? What do customers and suppliers seem to want?
- What are the driving factors for the business? What are the goals of the firm? E-Business must support these goals.
- Where does the firm make money? You do not want the project to interfere with this; you want to support making more money.
- Where does it lose money? Answering this and the previous question can help you identify E-Business opportunities.
- What e-critical activities should you consider at all? What supportive business activities should you examine?

Answering these questions correctly will help you select activities that have impact through E-Business implementation. Choosing the most appropriate business activities is crucial to your success, because it will be difficult, if not impossible, to change them later.

With the general purpose being to understand the business, find more specific objectives that can be linked to business activities. Your E-Business effort will then support these. For example, if you want increased profitability in general, identify the areas of the business in which this is possible. Going beyond this, seek to identify groups of business activities that will serve as candidates for E-Business projects. You will look for groups because most are interwoven and usually revolve around a major business activity that crosses multiple functions and divisions or departments. For example, setting up ordering for E-Business without considering fulfillment and returns does not make much sense.

MILESTONES

The major end products are business objectives and issues that represent a major focus of the business and comparison tables that tie the information together as well as activities that support these business goals. This applies to new startup companies because you then concentrate on your planned business. Because E-Business implementation can be both time consuming and expensive, the more narrow you can define potential business activities, the easier the implementation of E-Business will be.

METHODS FOR E-BUSINESS

STEP 1: UNDERSTAND THE BASIC BUSINESS

Read up on your organization in magazines, periodicals, and the financial press. E-Business at its core must be aligned with the basic objectives of the enterprise. What do outsiders say about your company? What do they list as its strengths and weaknesses? Who do they mention as competitors? What challenges does the firm face? What is said about your firm now as compared with two or three years ago? Use the web and CD-ROM databases in libraries to search for articles. Search for chat rooms and forums where customers mention things about the company and its products and services.

Consider the politics of the business. Where is the center of power? What recent organizational changes have taken place? How secure are the people in power? To what extent does their management interest lie in specific business activities? Why are the answers to these questions important? You cannot approach E-Business casually. It is a long-term proposition and not a casual whim to be played with by a transitory manager.

Look to see where customers or suppliers touch the organization and business activities most frequently. This may not be obvious. In the case of Ricker Catalogs, there were a large number of contacts from customers related to returns of merchandise that Ricker did not produce themselves. This might be a target area for E-Business.

Examine stable areas of an organization. Why are they stable? Is it because their transactions and workload are stable? Is the stability hiding inefficiencies? Hidden issues may lie underneath the stability. If an area appears stable after you have considered it in depth, move on to other business activities. For Abacus Energy the business activities relating to contracting and outsourcing were stable.

Culture plays a strong role in business. Some questions you will try to answer are the following:

- Does management encourage and reward teamwork? E-Business implementation is a team project. If the culture supports a mentality of every person for him- or herself, you will have another barrier to overcome.
- Are problems and situations openly addressed? The answer to this will point to the appropriate manner in which issues should be addressed during the E-Business project.
- Are employees encouraged to develop new ideas and to think for themselves? If there are two things that E-Business demands, they are the creativity of employees and their participation.
- Does management empower employees to make decisions and suggestions? If the answer is affirmative, it is more likely that E-Business ideas will be contributed by the staff involved in the work.
- What happens if people take risk? Is there any recognition or punishment? General work improvement involves political and workflow risk— E-Business even more so. If risk-taking is discouraged, the odds of success are lessened.
- Are people open with information or do they try to keep their knowledge to themselves? E-Business projects depend on the sharing of information and an open approach.

Remember that your goal is not to change the culture of the organization; it is to learn what you must face at the start of the project.

STEP 2: IDENTIFY PREVIOUS ATTEMPTS AT IMPROVEMENT

This may seem irrelevant to E-Business, but you want to learn from the past for the future. Tied to culture is the desire of the organization to improve itself. Over the past 20 years, many management improvement approaches have been tried in organizations. In addition, your firm may have a faltering or only partially successful e-commerce effort. The following is a partial listing of sample management approaches.

- *Vision of the organization.* This is a popular device to get a common framework and understanding of the focus of the organization. If there is a vision, it is tangible or fuzzy? Could it fit a hundred other organizations in different industries? Does the vision reflect reality? Do actions support the vision, or does the vision simply hang on a nice frame in the visitor entrance? For E-Business the vision must be more precise than just, "Establish an E-Business presence." Where does the company want to go with E-Business?
- *Analysis of the industry and competition.* This is an increasingly popular method. Talk to the people who perform this analysis. Find out if they are satisfied with the actions taken. Ask whether the analysis translated into action or remained passive information. Try to get leads for E-Business activities.
- *Alliances with customer and suppliers.* There may be existing automated relationships with customers and suppliers. Some companies have found such alliances beneficial in terms of quality improvement, electronic linkage, product design, and industrial practice. Finding out about alliances may provide clues to which business activities are the most important. You will also need this information as you begin to assess the impact of E-Business on these alliances.
- *Benchmarking.* Does your firm do benchmarking? Does it look for good practices and then fit them into the organization? Does the firm benchmark, and then do nothing with the results? Benchmarking will help in measurement of business activities and can assist in identifying critical business activities. There also may be comparative studies of web sites in terms of number of visitors, features of the web sites, etc.
- *Organization streamlining.* Has your firm attempted organization change without improving the business activities themselves? Have management layers been reduced? To find out, dig out old organization charts and compare them to the current version. This will provide valuable background information for the E-Business implementation. If there has not been any streamlining, for example, then this may be a necessary ingredient to implement E-Business for efficiency.
- *Assessing management's commitment.* Look at what the organization has attempted in terms of the following:
  — Using the value chain
  — Carrying out quality management programs
  — Retreating to core competencies
  — Undertaking efforts to reduce cycle time

— Empowering employees and organizing them into groups

— Embarking on new technology

Look at management's response and attitude toward these techniques. What was the level of commitment? Did failure make it more reluctant to try new methods? This is the reality with many ideas that are oversold and overpromoted. Expectations are unrealistic. Alternatively, did management start and then stop with these techniques. This indicates that you may face problems in E-Business if management changes its views.

Ascertain the areas in which management is willing to make changes. Alternatively, you might ask, "If you had the money and the time, what would you change?" or "What would be the E-Business target from a business view?"

Ask managers what they would like the customer or supplier of the future to look like in terms of geography, demographics, and other criteria. Will E-Business implementation move them in this direction? In one retail firm, the management wanted to address female, high-income customers. At the same time, they wanted E-Business. The E-Business project focused on attracting moderate-income males. Can you guess what happened?

STEP 3: DETERMINE HOW YOUR COMPANY ADDRESSES INNOVATION

E-Business is innovation with a capital "I." Organizations approach methods differently. Some are "innovators." They analyze a method, and then try to implement it and achieve results. "Followers" attempt to implement changes but may not have the will to follow through for major benefit. They may achieve some benefit and then stop.

"Dabblers" make up a third group. A dabbler investigates many methods, but they never carry out change. Dabblers sometimes initiate pilot projects, but never reach full implementation. A fourth category is "resisters." Resisters are led by people who distrust methods. They watch methods come and go and become cynical. They concentrate on basics.

E-Business projects can work easily with three of the groups. The fit with innovators is obvious. With followers, it will be the job of the project team to carry out the project and get benefits. There will be less management support. Resisters are an interesting group. Because they realize the importance of the basic business, they are more receptive to change.

The most difficult organization type is the dabbler. Often, dabblers do not take E-Business seriously until they feel threatened. They require a trigger to activate a serious attitude toward change. It is often best to explain first what will happen if they continue their current path. If they are still not serious, consider ending the project.

E-Business implementation focuses on business activities; it is not anchored solely in the software, infrastructure, the technology, or the organization. In E-Business you have to consider both current and new types of workflow and transactions.

STEP 4: DETERMINE YOUR BUSINESS STRATEGY FOR E-BUSINESS

You will refine this as you go. However, you should begin early. Here are some basic questions you should answer.

- What is your current value chain and what will be the value chain with the web?
- How will E-Business impact the value chain?
- What new business areas could you enter with E-Business?
- How can E-Business be employed to challenge the common means of doing business?
- What new value will E-Business provide to your customers or suppliers?
- What is the vision for E-Business? How will it be communicated?
- Is senior management aware of the threats offered by the web?

Posing and then answering these will raise the level of awareness of E-Business and its implications for your firm.

STEP 5: COLLECT INFORMATION

First, gather as much data as you can by passive means. Utilize organization charts, vision statements, long-range plans, policies, and procedures, and some external information on the company and industry. Passive information is readily available with little effort. Use the information to help generate interview questions. Try to develop some "strawman" statements related to the questions in Step 4.

Data collection should not take a lot of time. The project clock is ticking. You will have an opportunity later to collect detailed information. Some who prefer detail find this frustrating, but it is in this step that you get the big picture. You will relate everything that follows to the understanding of the organization obtained in this step. This information will provide suggestions for business activities. It will aid you in Action 5, when you collect information for business activities, infrastructure, and organizations.

STEP 6: IDENTIFY CANDIDATE E-BUSINESS ACTIVITIES

You have gained an understanding of the business. Now define candidate business activities for later evaluation and selection for E-Business. Your goal is to identify several candidate activities. An organization might have more than 250 business activities, depending on the definition of their scope, making it important to narrow the field of candidates. You want to choose 10–15 business activities because it is easier to choose at this stage than to add in the future and politically you must demonstrate a comprehensive approach.

To begin to identify and narrow the field of business activities to consider, address the questions listed below. Note that there are questions about the business activities themselves as well as E-Business because you need to understand the cleanup effort that will be required.

- Which business activities seem to be critical for E-Business in your firm?
- How do the activities interrelate with each other?
- Which business activities appear to cause management problems?
- Which organizations appear to cause management problems?
- Which activities appear to be important in generating revenue or consuming large quantities of resources?
- Which activities cross the most organizations?
- Which activities have not been analyzed or modernized?
- Which activities appear to have the highest cost in terms of people and infrastructure?
- Which business activities entail the highest volume of work? Which have the most transactions?
- Which business activities require the most reworking and produce the highest error rates?
- Which are the activities about which you know the least?

Ask the same questions about the computer systems and infrastructure that support the business activities.

Consider the obvious business activities that interface with suppliers and customers as good candidates for E-Business. Customer and supplier transactions are external so that they tend to link directly to revenue and cost.

The fact that a big consulting or accounting firm has analyzed a specific business activity does not mean it is a poor candidate. A previous analysis that indicated that an area is not fruitful does not mean that you should avoid it. Perhaps the previous study did not focus on improvement. The analysts may have had a different point of view. Enough time may have passed to merit reconsideration.

Figure 3.1 lists some of the business activities for the four example firms. Note that these are just some of the major business activities. For each firm there were additional supportive business activities.

Technology and Infrastructure

As another approach, look for business activities addressed by specific technologies. For example, consider those that can be addressed by intranets and extranets as part of E-Business. In insurance this could be linking outside claims adjusters to the insurance company through the Internet. Electronic data interchange (EDI) is the electronic transfer of structured information between organizations. EDI was a predecessor to e-commerce. EDI can facilitate ordering, invoicing, shipping, and receiving. Technologies and their producers obviously identify and market their products to specific applications.

Interview Maps

You also may want to map the people interviewed (rows) versus business activities (columns). Place an "X" in a column if that person mentioned the corresponding business activity as a concern. If you have a row with no entries, ask why this is so. You also could sort the rows by management levels as well as by organization. Interviews result in notes and understanding. Interview maps help to integrate and summarize the findings of the interviews. Be careful during interviews to push E-Business. This will come later.

Literature and the Web

Do not ignore the printed literature. Literature is another source of business activity information. Look in magazines and other printed material for information about leaders in your industry, which activities are regarded as excellent, and what other companies have improved. Do the same using the web. There are over 100 magazines available on-line through the web internationally. Bookmark these and then organize them. You will want to review these on a recurring basis. What are some search items for the web? Some suitable ones are e-commerce in your industry segment, e-commerce software, e-commerce magazines, and information on your industry segment.

- **Ricker Catalogs**
  — Contacts with vendors to supply products and text and photos for the web
  — Establishment and maintenance of the web content
  — Competitive assessment of other web sites
  — Marketing and sales approach for the web
  — Marketing data analysis related to products, customers, and sales

Figure 3.1 Business Activities for Example Firms

— Order processing
— Credit card processing
— Handling of back orders
— Handling of cancellations and refunds
— Handling of returned items
— Customer service of web customers
— Fulfillment of orders
— Maintenance of the web software and support for enhancements to support promotions and discounts
— Monitoring and measuring use of the web
- **Marathon Manufacturing**
  — Sales representative approach to contact smaller firms
  — Catalog content setup and maintenance
  — Lessons learned and guidelines setup and maintenance on the web
  — Order processing on the web
  — Wizard software for generating bids
  — Shipping processes
  — Inventory policies and procedures
  — Manufacturing order priorities and management
  — Management reporting
  — Marketing analysis
- **Crawford Bank**
  — Establishment and maintenance of the web site
  — Competitive assessment and monitoring of other web sites
  — Loan application on the web
  — Pricing and profitability analysis of web business
  — Loan processing
  — Customer service
  — Loan payments
  — Collection activity
  — Assessment of additional products
- **Abacus Energy**
  — Collection of internal requirements for bidding
  — Preparation of bidding materials
  — Establishment and maintenance of the bidders' list
  — Dissemination of requests for proposals
  — Handling questions at bidders' conferences and from bidders
  — Evaluation of proposals
  — Contracting and negotiation of the contract
  — Coordination of the start of work or delivery of goods
  — Handling disputes between internal departments and vendors
  — Defining modifications to contracts
  — Handling modifications and extensions to contracts
  — Maintaining the web services
  — Monitoring and measuring use of the web

Figure 3.1 continued

Step 7: Summarize the Chosen Business Activities

You now have a list of individual business activities to consider. To understand these more fully, prepare a brief summary table. In the first column list business activities. In the second column place characteristics of the activity, and in the third list characteristics of the systems and infrastructure. An example for Ricker Catalogs appears in Figure 3.2.

When you look at this table for Ricker Catalogs, you can see the daunting challenges they face. The other firms faced similar challenges. And, keep in mind, all of these had working business activities and systems that made money.

Creating such a table has several benefits, beyond support of the grouping of business activities. First, people can review the table quickly. Second, it provides a basic understanding of activities at a high level. Third, the table will eventually help you select and market the chosen activity group.

Step 8: Build Comparison Tables

Comparison tables in this step relate to business factors and activities. Building these tables allows you to analyze the activities prior to a detailed evaluation and selection. The tables serve to validate the identified activities as well as the business objectives. They also help to explain what you are doing by displaying it in an organized method. These tables appear quantitative but are actually quite subjective. Enter a number from 1 to 5, depending on the degree to which the column is an issue with respect to the row (1 is lowest; 5 is highest). For example, if the entry in the cell defined by row A and column 1 is low, column 1 is not an issue of importance for row A. If the entry is high, the E-Business project should pay some attention to it. From the information gathered, you can build the following tables:

- **Business objectives versus organization**
  The rows are business objectives as defined in the data collected. The entry in the table is the degree to which the organization supports the objective. Organization includes the major divisions and departments. Put in both the current traditional business objectives and E-Business objectives.
- **Business objectives versus the identified business activities**
  The rows are business objectives; the columns are the business activities that you have identified. The entry is the degree to which the activity impacts the objective.
- **Business objectives versus infrastructure elements**
  The rows are business objectives; the columns are elements of infrastructure. The entry indicates the degree to which the infrastructure contributes to the business objective.

Business Activities	Characteristics of Activities	Systems and Infra-structure Characteristics
Contacts with vendors to supply products and text and photos for the web	Modification of current business activities needed	Mostly manual effort; labor intensive
Establishment and maintenance of the web content	New	System needed
Competitive assessment of other web sites	Informal today; needs to be formal	Manual
Marketing and sales for the web	No formal processes exist; can salvage from catalog business activities	Lack of automated tracking
Marketing data analysis related to products, customers, and sales	Current activity will have to be expanded	Some statistical analysis tools
Order processing	Telephone call center	Automate for the web
Credit card processing	In place today	Interfaces are needed
Handling of back orders	Current approach is labor intensive	Some automated support
Handling of cancellations and refunds	Current approach is labor intensive	Some automated support
Handling of returned items	Current approach is labor intensive	Some automated support
Customer service of web customers	In place	System exists today
Fulfillment of orders	Backorders are common; need to reduce	System exists today
Maintenance of the web software and support for enhancements to support promotions and discounts	New	New system needed
Monitoring and measuring use of the web	New	New system needed

Figure 3.2   Ricker Catalogs Business Activity Characteristics

- **Business objectives versus issues/opportunities**
  In the information you have gathered are certain key issues. Include both
  E-Business and general issues. This table reveals the degree to which resolving the issue or taking advantage of the opportunity supports the business objective.

Note that E-Business is not mentioned in these tables. E-Business will be reflected in the table entries for business activities. Here you are after general business summaries through the tables.

E-BUSINESS EXAMPLES

RICKER CATALOGS

Ricker Catalogs was used as an example in Figure 3.2. Preparing the tables
provided management with a real eye opener. They quickly eliminated customer
service as a activity to touch since their E-Business plate was full with the catalog,
lining up products, and order processing.

MARATHON MANUFACTURING

Marathon proceeded to develop the tables. Remember that they had time on
their side. They did not eliminate any of their business activities. Instead, they
encouraged their employees to add to the list of activities. That is how the lessons
learned and wizard for bidding on jobs appeared on the list. They involved many
employees in the review of the tables. Participation turned out to be more effective
than training since people were more directly involved.

ABACUS ENERGY

Abacus Energy's efforts were more narrowly focused. The tables helped management and senior employees to see the shortcomings of the current activities.
Abacus also involved several respected vendors that they had done business with
over the past 20 years.

CRAWFORD BANK

Crawford Bank scanned web sites to develop their initial list of business activities. Then they added more internal activities. At the review sessions, they found
many managers who were clueless about the web. This triggered substantial train-

ing for the managers in the form of orientation. They also hired a consultant to develop the tables and do some of the analysis.

E-BUSINESS LESSONS LEARNED

You will likely be providing management with informal presentations on what you have collected so far. Show the comparison tables to get management's reaction and feedback. This accomplishes several goals. First, you are establishing a pattern of looking at concrete information in the form of tables. Second, it allows management to get involved early in the lead-up to the E-Business projects.

What should you expect from top management? At this point they should be able to give you feedback on your initial ideas through the interviews and tables. However, don't commit to specific business activities at this point. You do not have enough specific information. A false start with a direction change can be deadly. Watch for managers who are enemies of change and E-Business to give out false signs and set traps.

By considering the issues that are raised in this action, you can start to make out potential substantial benefits. However, if you have to discuss costs and benefits, do so by pointing to the *potential* without getting pinned down to specific figures. Indicate that more data will become available in subsequent steps.

WHAT TO DO NEXT

1. Identify three business objectives applicable to business activities for E-Business. For each, indicate which goals are relevant from the following list. Keep the wording of the objective simple.

1. Increase sales through new products
2. Increase sales through additional customers
3. Increase sales through repeat business
4. Establish closer ties with suppliers
5. Handle higher volume of work with same resources
6. Reduce overall operating costs
7. Reduce staffing
8. Reduce infrastructure and capital costs
9. Streamline business activities
10. Realign power between the enterprise and the divisions

Objectives:

1. _____

Goals: _____

2. _____

Goals: _____

3. _____

Goals: _____

2. Identify 10 business issues that apply to your business activities. For each, indicate what will likely or possibly occur if the issue is not addressed. These business issues will be related to business activity selection and other parts of E-Business implementation later.

Issue Implications

1. _____ _____

2. _____ _____

3. _____ _____

4. _____ _____

5. _____ _____

6. _____ _____

7. _____ _____

8. _____ _____

9. _____ _____

10. _____ _____

3. Construct a table in which rows are organizations or divisions and columns are key business activities. In the table indicate the importance of the activity to the business unit on a scale of 1 to 5 (1 is not important; 5 is very important).

Business Activities

Organization				

4. Construct a table in which rows are elements of infrastructure and the columns are key business activities. Indicate the importance of the infrastructure to the activity on a scale of 1 to 5 (1 is not important; 5 is very important).

Business Activity

Infrastructure				

5. Relate business activities to each other in the following table. Rows and columns are activities. The entries in the table can be any combination of the following: I, share infrastructure; O, share organization; IN, interface with each other; C, share customers; S, share suppliers; and SY, share systems.

Business Activity

Business Activity				

6. Relate the business objectives to the business activities in the following table. Each entry is 1 to 5 (1 is not related; 5 is highly related).

Business Activity

Bus. Objective				

Chapter 4

Action 2: Select the Activities for E-Business

INTRODUCTION

E-Business success comes from making the best selection of business activities for E-Business as well as implementation. There will have to be multiple related activities for E-Business because of the need to provide a complete set of services to customers or sufficient benefits to warrant supplier involvement. In any organization there are a large number of business activities that yield many different combinations of activity groups.

In this action, you will carry out a grouping method whereby you define clusters of related activities that relate to E-Business. You will then choose one group for E-Business implementation. Beyond E-Business impact, you will be considering selection based on factors such as politics, work stability, investment required, potential benefits, and long-term direction. You do not need to avoid taking on activities that are politically active and sensitive—as long as you know about this in advance.

You begin by identifying one or more key business activities. These are centerpiece candidates. Then you add related activities to form a candidate for a group. In E-Business there are typically several potential candidates to center on for a group. For Marathon Manufacturing, these were ordering, the product catalog, order fulfillment, and lessons learned/guidelines about products. Managers at Marathon knew that they had to deal with all of these for E-Business. Marathon had limited resources but no real time pressure since no one in their limited industry segment was doing much with E-Business. For Marathon the issue was long-term competitive advantage, not just getting out with some standard web site that everyone would copy.

For Ricker Catalogs there was severe time pressure. There was little flexibility. If they picked activities that might be needed later for E-Business, they might not reach a competitive position fast enough. Alternatives for Ricker included ordering, customer service, returns, fulfillment, and the catalog. The question was where to start.

In the case of Abacus Energy, there was a drive to outsource. Outsourcing activity was swamping the contracting and purchasing group. Complaints were rolling in from both internal departments and external vendors about the business activities. The question was how comprehensive the initial E-Business effort should be.

Crawford Bank faced a challenge at the start. Crawford was put in a reactive mode when several banks offered installment loans on the web at lower rates than Crawford. Crawford knew that they had to respond. To answer the challenge decisively, Crawford also knew that they would have to streamline their internal work and transactions to drive costs down (to yield competitive rates) and to provide more complete customer service.

You will need criteria and trade-off analyses to choose among the possible activity groups. You want to select a activity group that fits with your E-Business situation. Each of the four companies faced different challenges. In fact, as you can see above, Marathon and Ricker had two opposite situations. You will also want to select a group of activities in which the following is true:

- Sufficient commonality exists between activities in the group so that they make sense from a customer or supplier point of view.
- Implementation can occur in a reasonable time to respond to E-Business pressures.
- The new E-Business transactions and workflow will be stable enough to provide benefits for both E-Business and traditional business.
- Work with the first group of business activities paves the way for addressing another group. This supports a wave approach to E-Business implementation and growth.

Recall that you have chosen a number of activity candidates as centerposts for E-Business. There are a number of alternative ways to group activities. You can group around activities that involve risk and require improvement due to their condition. This is viable if you have a longer elapsed time for E-Business implementation. You can group by organization (such as marketing or customer service). This is not really suitable to E-Business since E-Business activities cross organizations. You can group by function such as ordering or customer service. This is probably a good method to get E-Business implemented quickly. For each approach you can select several candidates as centerposts. Then you can choose any number of related activities to form a candidate group.

Keep in mind that if you leave one or more activities untouched, management may feel exposed and vulnerable. They could rush to implement "Quick Hit" E-Business processes and then find that they have limited flexibility. All of their effort could be undone due to the risk and issues in implementing other parts of E-Business.

A recommended approach for E-Business is to proceed by exclusion. Determine what to recommend for the business activities *not* selected. Are they next in line? Are they to be untouched? If so, for how long? How do they fit into the picture? This differs from other methods that concentrate on the winners of the analysis. You are concerned with the "activity portfolio"—a group of activities.

MILESTONES

The basic end product is your assessment of the business activities in terms of groups. You will recommend a specific group of activities for E-Business implementation. You will also indicate why other groups and activities are less attractive for improvement at this time. Your selection will be supported by tables and graphs.

Your political milestone is the concurrence and agreement across management and senior staff that the activity group selection for E-Business was correct. Look at the downside. If you do not have consensus or near consensus, you could have the activities selected change based on some market or competitive condition. Then your work is undone and you have to start over. You risk not implementing anything—a dismal prospect.

METHODS FOR E-BUSINESS

STEP 1: DEFINE THE BOUNDARIES AND INTERFACES OF BUSINESS ACTIVITIES

Look at the list of the business activities that you generated in Action 1. Where does one activity end and the next begin? How do the business activities tie together? The answers to these questions help to determine whether an activity belongs to a group or not. This is critical in E-Business because if you fail to include linking activities in the same group, then you have to implement manual, stopgap measures in the middle. Some ways two activities interface are as follows:

- One passes work instructions to another. One activity depends on another for procedures and information. This is the strongest type of linkage. An example of this is order processing and fulfillment of orders.
- The two share of information. One activity feeds information to another.

The second activity is therefore highly dependent on the first. An example of this is order processing and customer service.

- One depends upon the completion of the other. An activity cannot begin until the previous activity is completed. An example of this is order processing and credit card billing.

In E-Business you first position the catalog on the web. Customers visit your web site and place orders. The orders are then entered. Order processing occurs along with credit card billing. Fulfillment of the order then happens. Later activities are cancellations, returns of merchandise, and handling back orders. On top of these you have marketing activities related to discounts and promotions.

Many other variations are made possible by combining any two of the factors. If there is a strong relationship between the content of one activity and another, consider those activities together.

Two activities do not have to interface directly. They might relate to each other indirectly through infrastructure, organization, and resources. You may not think that facilities and infrastructure are important to E-Business. But remember that you will be changing the internal non-E-Business workflow and procedures at the same time that you are implementing E-Business. Some examples are as follows:

- They share hardware and a network. Both operate on the same infrastructure. Changing one could impact the performance of the other activity. If you change an activity to support increased network traffic, you impact network performance. If you increase the number of users or extent of use, you impact traffic.
- They share facilities. Two activities that inhabit the same space will impact each other after one of the activities changes.
- They share equipment. The activities may conflict over specific equipment. Different equipment and levels of equipment may be necessary following change. If the same staff is to operate both sets of equipment, problems will arise.
- The same people perform both today. Changing one activity may be counterproductive to the other activity.
- They impact the same supplier or customer. Changing one activity may mean that the external organization now must work with two entirely different and incompatible business activities.

When you consider a specific business activity, identify other activities that are *tightly linked,* meaning that any significant change (requiring a project and management approval) will affect the performance, cost, or requirements of the other. Activities that are *loosely linked* should be considered separately. Loosely linked might mean, for example, that while different groups perform each activity, both groups report to the same manager. In E-Business, ordering and credit card authorization are tightly linked. However, the link between ordering and accounting after the order are loosely linked.

You might also consider a business activity by examining the impact of changes made to see if there is any effect. For example, suppose you eliminate activity A. Will activity B be affected? If you move activity A outside (outsourcing) or change organization or infrastructure (rightsizing), what will be the impact on B? Assessing some potential impacts helps to determine groupings.

For Abacus Energy generating requests for proposals or bids is tightly linked with contract negotiation. In turn, both are more loosely linked to contract extensions and contract problem resolution.

STEP 2: GROUP THE ACTIVITIES

Using the list of activities, generate an activity group. Some criteria for generating a group include the following:

- Customer-based. Group all activities that touch the customer. For Marathon this could be all sales and servicing. For Ricker this encompasses all activities that touch the customer. In general retailing, you could consider point-of-sale, bar coding, and other activities together in a retail store. In banking, you could group all lending transactions such as loan sales, promotions, loan applications, application processing, billing, customer service, and collections. This grouping has tangible criteria, but critical internal activities may be ignored.
- Supplier-based. Group all activities that interact with the same suppliers. These activities could be those associated with ordering, shipping, and receiving.
- Function-based. Group all activities that perform the same function. This is one of the safest groupings, but it will probably make the project more complex by spanning multiple organizations. In banking, this could be all servicing activities. In insurance, this could be all claims processing through electronic means.
- Organization-based. Group all activities that the organization operates and manages. In banking, this could include all credit card transactions. A bank, for example, might want to establish a separate entity for E-Business. Note that the placement of functions in the current organization may be the result of politics rather than logic.
- Technology-based. Group all activities that use the same technology. The same hardware, software, or network can serve as the basis for technology-based activities.
- General manager–based. Group all activities under the direction of a high-

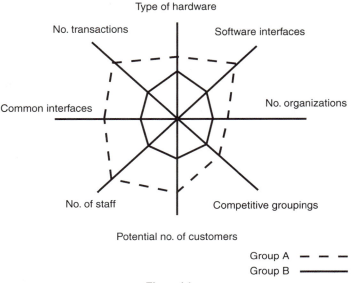

Figure 4.1

level manager. Most organizations have many layers of management. Thus, a high-level manager may oversee a number of organizations. Note that this grouping leaves you open to the charge that the project is too narrow.

- Financial performance–based. Group all activities that lose money or incur too high a cost. In the banking example, these could be activities that have substantial losses or generate profits.
- Business objective–based. Group activities that support a specific objective. Be careful not to use too general a business objective.
- Competition-based. Group activities that are keys to competitor success.
- Issue-based. Group activities that generate the most complaints.

You will want to consider all of these for E-Business. You can sometimes use a graph (as in Figure 4.1) to clarify the reasoning behind the grouping of activities. Note that while you may not have precise numbers for each axis, you can estimate. Figure 4.1 shows two group alternatives (A and B). The eight axes cover these dimensions of improvement:

- Number of organizations involved
- Common interfaces with customers or suppliers
- Competition groupings of activities for E-Business
- Number of users in departments affected

- Size of the potential customer audience
- Number of transactions
- Variety and number of types of hardware and network components
- Number of software interfaces

A major business activity serves as the center for both groups. Group B is more narrowly focused than Group A. It involves fewer organizations, transactions, and technology platforms and systems than Group A. Group B is likely to deliver larger benefits in the long term than Group A. As well, Group A involves more risk and takes more elapsed time to implement.

With more people, organizations, and managers involved, there is more complexity than is found in Group B. Keep this diagram to show comparisons between groups. Alternatively, you can use a bar chart in which each group has a set of bars, and each bar corresponds to a specific axis.

Another guideline for constructing a group is to include some simple, related activities in which rapid benefits may be demonstrated through prototyping, piloting, and fast rollout.

To the group of basic activities add smaller activities within departments and general purpose activities. These will support the core E-Business activities. Some examples of subsidiary activities are the following:

- Support for letter generation or faxing from output of the primary activities. For E-Business this could include order conformation.
- PC-based activities that use information generated by a primary activity. This could include statistical analysis of sales data on the web.
- Electronic forms or electronic mail that link to an activity.
- Groupware that performs a supporting role for an activity.
- Manual steps and business activities that are extensions of an activity.

These secondary activities are important because they fill out the picture and contribute to the overall benefits of E-Business implementation. Ignoring them could, perhaps, lead to a new business activity that is worse than the replaced one because it is more awkward and cumbersome. By including them, you might achieve earlier improvements.

Potential Problem: Too Many Activities in a Group

If too many activities are involved, infrastructure changes may bog down the project. You may be slowed down if too many organizations are involved. The scope of the E-Business is too wide. Look over your choices again. It is inefficient to group activities that are more different than alike. If the activity group is still too large, stretch out the project and raise the level of risk and exposure. Also increase the number and extent of reviews, status meetings, and coordination efforts.

STEP 3: BUILD COMPARISON TABLES

Activity Group versus Activity Group

Be able to discuss how activity groups are compatible and supportive of each other. Two groups are supportive if one naturally leads to the other. Groups are not supportive if they require different organization or infrastructure. Two groups are neutral if they have little in common. Construct a table with entries on a scale of 1 to 5 (1 is incompatible; 5 is highly compatible). Later, when groups are ranked, and sequencing is established, this table will provide assistance and support.

Activity Group versus Customer or Supplier Activities

The rows are the activity groups. The columns are activities that you expect customers or suppliers to perform in E-Business. The entry is 1 to 5, based on the degree to which the activity group supports the customer or supplier activity.

Activity Group versus Technologies

The rows represent the activity groups. The columns stand for technologies that you currently use or have a strong desire to use in the future. The entry is 1 to 5, based on the degree to which the technology is involved in the activity group (1 is not at all; 5 is very dependent). You are concerned with the technologies on which the activities will depend when E-Business implementation is completed.

This table determines similarities between groups in terms of technologies. If you were later to select two groups that used incompatible or separate technologies, availability of people and money might prevent you from completing the task.

Activity Group versus Infrastructure

The rows are activity groups; the columns are key infrastructure elements. The entry is 1 to 5 based on the degree the activity group will depend on the specific infrastructure element (1 is irrelevant; 5 is critical).

This table is important because it reveals which activity groups would benefit from investment in a specific infrastructure component. Where money is limited, this can be a determining factor in sequencing activity groups behind a leader.

Activity Group versus Organization

The rows are as before; the columns are the major organizations involved. This table shows patterns. All activity groups might fall in the same organizations.

There would be several possible explanations for this. Perhaps you have not done your homework and should return for more activities. Alternately, one organization might be in disarray.

How do you obtain data to fill in these tables when you have not yet performed detailed data collection and analysis on each activity group? Base the tables on the interviews and data collected in Action 1. In a perfect world, you would analyze all groups and then do the selection. Time, resources, and management pressure do not permit such luxuries.

The tables define the groups and set priorities between groups in a way that is logical, complete, and credible. The columns assist in your evaluation.

STEP 4: DETERMINE TWO WINNING GROUPS

Why do you want two winning or finalist groups? You want to give management alternatives. In the previous action, you constructed tables of business objectives versus business activities, customers/suppliers, organizations, infrastructure, and issues. The issues table points to which objectives are important. The organizations and infrastructure tables indicate which are critical to key business objectives. They help in your later analysis. Here most of your focus is on the activity table. This table indicates the degree to which the key activity in each group supports the business objective.

Evaluation Criteria

Define the criteria to use for performance—internal, external, and the E-Business project itself. A useful and fast technique is the development of a straw-man set of criteria (such as the ones listed above). After seeing the graphs, people may then suggest additions and changes.

If activity groups are different, instead of the traditional approach, take the following steps:

- Develop additional comparison tables for each group. These include activity group versus activity group, activity group versus technologies, activity group versus infrastructure, and activity group versus organizations. These additional tables, along with those from Action 1, will support the analysis.
- Determine conditions under which each alternative would win in the competition. Review the tables from Action 1.
- Proceed by method of elimination. Eliminate all groups that do not strongly support major business objectives; those that remain will all meet a business requirement.
- Rank the remaining groups in each of the following categories on a scale of 1 to 5:

— **Revenue from E-Business**
  Revenue and cost. The degree to which changing the activity group will
  result in increased revenue and reduced costs (5 is great benefit).
  Competitive position. The impact of the activity changes on the competitive position of the business (5 is great benefit).
— **Internal**
  Organization. Impact of changes in the activities on the organization's
  structure and head count. Note that this overlaps savings (5 is great
  positive impact).
  Infrastructure. The effect of activity changes on the infrastructure in
  terms of making the activities more responsive and effective (5 is great
  impact).
— **External**
  Customer. The degree to which the activity changes improve and enhance the relationship with customers.
  Supplier. The degree to which the activity changes improve and enhance
  the relationships with suppliers.
— **Project**
  Risk. The overall risk in the project (inverse ranking here—5 is low
  risk).
  *Elapsed time and overall effort required to carry out the E-Business
  project.* Use inverse ranking here (5 is low effort and short time).

You could substitute different criteria, but try to cover the same four areas. You
are now able to construct a graph similar to that in Figure 4.1. The axes of the
graph are the criteria. Now proceed to these activities.

- Identify two different activity groups that are winners, depending on which
  criteria are important. Alternatives are significant. They offer a way of dealing with involvement, commitment, and politics.
- Develop a scenario or model of what the business would look like if the
  changes in each of the two activity groups were carried out separately. This
  model is a representation of how the new business activity would work. It
  does not include details of infrastructure or organization.

Identifying two activity groups gave Marathon management the opportunity
to discuss the trade-offs between short-term Quick Hit E-Business and strategic
E-Business.

Conduct the Evaluation

For each activity group, write a list of bulleted items that describe how you
think the activity group fares with specific criteria. This supports the defense of

your later recommendation. Next, assign the numeric value to each criterion. The first graph you will show is Figure 4.2. This shows the rankings for each of two activity groups A and B. This can sometimes be confusing due to overlap, so there is another chart, Figure 4.3, that combines all criteria in each area and assigns an overall rating. The analysis of these charts should indicate the two activity groups that are the most promising and indicate which is preferable.

STEP 5: DEFINE A FIRST SCENARIO

With these two activity groups identified, there is still further work. You want to be able to give a presentation and generate support. You need to show the audience how the new activities will really work after the completion of the project. If you give people an idea of how the implemented business activity will function, you will inspire belief in the benefits and analysis. To do this, create a scenario which is a model or description of how the new E-Business activities would work. You will refine the scenario created here in a later chapter, when the new business activities are defined.

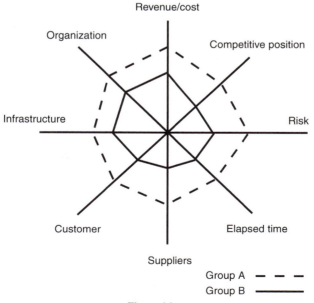

Figure 4.2

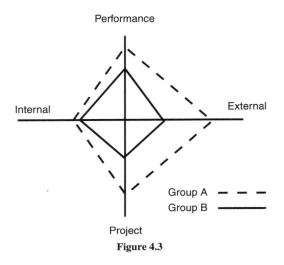

Figure 4.3

STEP 6: DOCUMENT THE SELECTION

You will probably have to develop a report to support your selection of the activity group. In some cases, you can manage to avoid this effort and launch the next stage of the E-Business project through marketing and explaining the comparison tables and the scenarios. However, if you must write it up, minimize the amount of text. Avoid technical terms. Stick to terms commonly used in the business. Develop your exhibits first. These should be comparison tables and the steps involved in the existing and new business activity. Also, develop the project plan for the next stage of effort. This material can enhance your presentation, helping to ensure that presentation and report are consistent and mutually supportive. This will also save time.

POTENTIAL MANAGEMENT ISSUES

You are likely to run into several management questions and issues. Some of these are the following:

- **Assurance of the choice of activity group.**
  Accepting that management may perceive risk and uncertainty in an E-Business project, it is natural and important to reassure them. In addition to providing them with the analysis results and the plan, indicate what will

happen if there is no action. (You will address this in the action on marketing.)

- **Concerns about activities that will not be improved.**
  What will happen to the other activities? It is wise to develop an overall strategy for them. You have looked closely at strategies for organization and infrastructure, while little effort has been spent looking at what makes money and incurs cost—the business activities. Placing a business activity in the context of an overall strategy is a good way to reassure people about the future of an activity. Included in such a strategy are replacement, improvement, and elimination.
- **Concerns about staff and organization during E-Business implementation.**
  Two actions should be taken to calm staff fears, which always arise. The first action is to have a project plan and documentation that focus on improvement. This reveals what you will be doing. Second, write your approach with a focus on benefits that relate to saving paper, simplification, and better control. This can be part of the plan.

Management will begin to get concerned about costs, schedules, and benefits. Discussing the benefits is possible from the comparison tables. To discuss costs, gather costs and schedules of similar projects in the past. These may not fit the current situation exactly, but they will give management an idea of relative size. Management support here can be measured in terms of their interest in the project. Is it growing? Do they want more detail?

How do you know if you are on the right track? Look at the activity groups you have identified. If you made changes to a group, how would the organization and activities benefit overall? If you deleted a specific activity from a group, what would be the impact?

Another test is to assess the level of interest in the project at the start of this step and at the end. The level should increase. It is in this step that you want to capture people's interest and imagination.

E-BUSINESS EXAMPLES

RICKER CATALOGS

Ricker knew the business activities to potentially center on were the catalog and the order processing. However, if they spent too much energy on the catalog, this would detract from other changes to fulfillment and customer service. So they chose order processing. Ricker's analysis led them to a Group B-type activity group.

MARATHON MANUFACTURING

Marathon had several alternatives. One was order processing. Another was to take a broader view of customer service and added value. They chose the second approach because they had the time to carry out a larger-scale implementation. Marathon's analysis led them to a Group A activity group.

ABACUS ENERGY

There were also several choices for Abacus. They could focus on the bidding and award of contracts or they could focus on contract setup or contract maintenance. They decided to start with the bidding activity. Their fit would be between Groups A and B.

CRAWFORD BANK

Crawford Bank had a dilemma. They could copy the competition and focus on loan sales and applications. Alternatively, they could take a broader view with the aim of eventually setting up more products and more extensive product support on the web. They opted for the second alternative—Group A.

E-BUSINESS LESSONS LEARNED

- **Always focus effort on e-critical business activities.**
  Activities with immediate management interest can consume your focus. It is also easy to be sucked into a technology pit. Keep your focus on the key activities relevant to E-Business—those that contribute to revenue or have major costs.
- **In E-Business do not be too centered on the straightforward customer activities.**
  Remember you are looking toward the long haul. You are going to be in E-Business for keeps.
- **Concentrate on business activities that cross organizations and are not within a department.**
  Almost all of your E-Business activities will cross departments. Within a department, activity changes tend to have fewer benefits (due to the restricted scope of the activities) and are less applicable to E-Business. Taking on an activity over several departments allows you to deal with a family

of business activities within those affected departments. Your work will have a larger impact.

- **Focus on business activities relevant to E-Business that are stable. Long-time existence usually means that improvements are possible.**
  E-Business is automated. If the underlying transactions and workflow are not stable, then you will be in trouble. Software change is not rapid and the environment is not forgiving. Stability helps in several ways. First, the activity has been performed for some time. If you improve it, the new activity will likely be there for some time so that you can recover the costs of your project. Second, the activity probably has ad hoc ways of handling problems and issues. These give you more improvement opportunities.
- **Focus on activities that contribute to revenue, as they are often more important than those that contribute to cost.**
  This is especially true with E-Business. Activities geared to revenue tend to be externally driven due to pressure from customers. Internal activities that are administrative are more difficult to resolve and less visible. Fixing workflow and procedures that relate to revenue will tend to show faster business results.

WHAT TO DO NEXT

1. Write down three critical business activities. For each, identify what other activities could be grouped with it, based on the criteria in the chapter. Indicate the reason for including each in the group. Also, indicate the strength of association on a scale of 1 to 5 (1 is weak affinity; 5 is strong affinity).

Key activity: _____

Related Activity Reason Association

1. _____ _____ _____

2. _____ _____ _____

3. _____ _____ _____

4. _____ _____ _____

2. Indicate the strength of the relationship between organizations and candidate activity groups using combinations of the following codes: O, organization owns the activity group; U, organization uses the results of the activities; I, organization is involved in the work; and N, not involved. This table helps to select the activity group and identify groups that impact multiple major divisions. Use the business activity versus organization table from Action 1.

Activity Group

Organization				

3. Construct a table of activity groups and systems and infrastructure in which the table entry is the extent to which the specific system or infrastructure element applies to the activity group on a scale of 1 to 5 (1 is no relation; 5 is very dependent). This table indicates the scope of the infrastructure and systems you may have to change in E-Business implementation.

System or Infrastructure Element

Activity Group				

4. Construct a graph like Figure 4.1 for your two main activity group candidates. Alternatively, you can construct a bar chart in which each activity group has a set of bars—one for each attribute.

5. Assess the degree of strength between the activities in a group. Construct a table in which the rows are the activities. The columns are organizations involved in the business activity, automated systems used in the activity, issues involved in the activity, condition of procedures and policies, volume of transactions, the network, and infrastructure.

6. Compare your top two activities by isolating the key activity in each group and comparing these with each other. Then correlate this with the group comparison of the previous steps.

Chapter 5

Action 3: Assess E-Business Trends and Competition

INTRODUCTION

Although much of your E-Business implementation effort will draw heavily on a company's internal business activities, infrastructure, and organization, you must be aware of what the industry and your competition are doing. You do not want to find out, after spending a lot of money implementing E-Business, that your competitors have already surpassed what you are about to roll out. As an example of the importance of market intelligence, a bookseller was going to implement a web site. Although they had surveyed competitors, they did not think about what these firms would be doing next. By the time they deployed their web site, it was already obsolete.

Another example involved the founder of a major microcomputer firm who invested some of his money in a company that made a programmable remote control device. His ego was such that he did not even consider the marketplace. After spending millions of dollars and building a prototype, he and his staff visited an electronics show. To their astonishment, they found many remote controls more sophisticated than theirs available at the show. He closed the company shortly thereafter.

There are also proactive reasons for collecting information. First, you may get ideas on how to do something better. Second, you might find out how to implement a new technology with less effort. Third, you can set standards that will have to be met. The changing nature of competition is another factor. Global competition has received much attention. You should know about new products, methods, and markets apart from your traditional base of business. Companies that do not have this knowledge sometimes find themselves locked into heavy competition.

BENCHMARKING

Benchmarking is an approach for evaluating products, services, and business activities of an organization. It can be internally or externally based. Many view benchmarking as a continuous approach that can improve business. This is definitely the case for E-Business. There are several types of benchmarking, including the following:

- *Type 1: Internal benchmarking.* If an organization has several divisions in different locations that do the same or similar work, internal benchmarks can usually be determined. For example, profitability of divisions is one financial benchmark. Internal benchmarking is useful, but it lacks comparison to outside standards. However, you will want to compare the basic activities that you will be changing for E-Business.
- *Type 2: Direct-competitor benchmarking.* Another type of benchmarking considers only those competitors that sell to the same market segments. Retail firms and computer firms are two examples. One problem with this type of benchmarking is that it can be difficult to collect data on competitors in an ethical manner. For E-Business you would examine their web site and look for information on the web about trends in the industry segment.
- *Type 3: Functional benchmarking.* Function is the basis for the third type of benchmarking. It entails comparing your company to the leaders of specific functions. For example, if you want to evaluate customer service, warehousing, and logistics, find the best firms in each area and compare yours to theirs. This is very useful in E-Business since you would want to consider the best in catalogs, the best ordering, and the best fulfillment.

Benchmarking can be done on a continuous basis or on a one-time basis. In E-Business implementation there is benefit in continuous benchmarking since other firms are changing their web sites dynamically in terms of look and feel, functions, and products and services. This sounds depressing in that it involves more work, but it is really positive because you have much more information available on competitors than would normally be the case.

Benchmarking can support E-Business implementation in many ways. It can provide input for a planning method; help to determine forecasting; and be used to compare products and business activities, derive new ideas, and set reasonable goals. It can also provide design ideas, product ideas, etc.

THE NATURE OF COMPETITION IN E-BUSINESS TODAY

Most large companies have full-time staff and outside consultants doing competitive and technology assessments for their own and related industries. Numerous surveys and magazines offer comparisons and examples. In a recent survey, more than 450 magazines were found to carry such comparisons.

Competition in E-Business is intense and heating up. There is information suggesting that once you capture a customer and they receive good service, they will return to the site rather than shopping around. E-Business, like regular business, thrives on repeat customers. Customers who visit a site several times feel more at home with the site in terms of navigation and content. The thrust of these comments serves to indicate that firms are driving to get to customers first.

Another competitive approach is to work on getting a much better web site than the competition. This then would distinguish your site from others—attracting more visitors. The problem here is to define the phrase "much better" in terms of the web. Better does not mean prettier; it often means greater functionality, more information, and ease of use. This approach also delays your entry into the marketplace.

In terms of competition, it seems that the best approach is to be somewhere in the middle. You do not rush out there with a trash site, nor do you work on it forever. Later, we will consider the appropriate E-Business implementation strategy—it will be in between these extremes.

Wal-Mart is considered by many to be an example of a successful firm that has leveraged its investment in its internal business to increase market share and profitability. Wal-Mart is somewhat unique because it addressed multiple business activities at the same time. It addressed warehousing, ordering, receiving, sales, stocking of inventory, supplier relations, and accounting. Wal-Mart's purpose was to accelerate the flow of goods through its retail channel to the customers. By buying more goods at a faster pace, it obtained better prices and had a greater impact on suppliers. By selling more goods, Wal-Mart increased sales per square foot. The result is not only greater profitability, but the ability for Wal-Mart to expand rapidly into new areas by using the infrastructure and organization associated with its new business activities.

Wal-Mart has now entered the web using the same successful approach. The strategy has been to link the stores and web to build cross-selling opportunities. This will be a good example to watch over the years because it will be a valuable example of how E-Business and standard business can exist side by side and be mutually supportive.

Many other retailers have studied Wal-Mart to attempt to learn from this example. These firms, however, are not always successful. One reason is reluctance to embrace so much change. For others, ingrained culture and rigid organizational structure prevent change. One firm lost more money after installing point-of-sale, electronic data interchange (EDI), and other technologies. The key is not technology but it is the business activity, and the firm did not change the business activity.

There are several lessons to be learned. First, you cannot ignore a successful company such as Wal-Mart if you are considering similar business activities even if you are in a different industry. Second, these leader firms set standards and patterns of business. To ignore them is to risk disaster. Third, it takes more than simply installing some changes and technology to achieve success in both regular business and E-Business.

COMMUNICATING FINDINGS

There is a tendency to tell people about your findings, especially when the data contradict a previous position. Do not be tempted. You are unlikely to be successful if you attempt to copy and apply what you find directly to you. Also, avoid applying a solution. Instead, start building the following:

- Web sites of other firms that may not be competitors, but offer similar products or functions
- Examples of success that have appeal and interest to people in your company
- Lessons learned about key issues relevant to your firm
- Statistical graphs and tables (even if they are only partially complete)

The primary objective in gathering external information is to obtain external information about the business activities you are considering or have selected. This will support your efforts in developing E-Business activities.

MILESTONES

Major end products of assessing the competition and industry are as follows:

- Comparative tables. Specific tables will provide a good summary for management.
- Graphs and tables on selected data. Use a few graphs or charts in your reports for clarity.
- Comments and highlights on business activities, issues, and use of technology. These comments should apply to the specific activities selected rather than to the business in general.
- Database of source information. This will be useful in the long term as you move through the organization with other E-Business projects.
- Files of information providing more in-depth information than the database.

Here are some categories of resources that will be useful.

- **People who have outside contacts.**
  Look to marketing, sales, and human resources. However, do not limit yourself to these departments. Consider the example of a major drugstore distributor. The company distributes drugs and other items to drugstores and retailers. The company managers decided to use employees as part of the intelligence and marketing effort. After the employees delivered the products, they reviewed the shelves not only for the company's products, but also for those of competitors. This information was sent back to the company. This effort contributed to the company's success. In E-Business that is the same approach that Marathon Manufacturing employed.
- **Employees who have reputations as gurus within the company.**
  These are people who know what is going on through reading and contacts. Identify these people early. Establish regular contact. In a bank, one collections supervisor belonged to several professional groups and had a wide range of contacts. He was not a star performer but he had a great deal of experience and was respected in the industry. Involving him in the project not only helped the project, but gave him the recognition that he deserved. In the case of Ricker Catalogs there were two employees who routinely tracked web catalog sites on their own. Getting them on board the E-Business project was very useful.
- **People who can help you do the legwork on the project.**
  Look for people who are inquisitive, are good problem solvers, and have outside interests. They may be avid readers, collect articles, and enjoy detective work and problem solving.
- **Suppliers and customers.**
  Build up relationships with suppliers and customers. They can provide information on the industry and competition. Even if the information is not detailed, you will get their perspective. Suppliers and customers will tell you how the firm and selected activities are viewed and what they like and dislike. Ricker Catalogs conducted some limited surveys of customers to see if they would shop on their web site. Abacus Energy surveyed their suppliers to determine which used EDI, e-commerce, and the web and Internet with their customers. Marathon conducted a survey of potential smaller businesses to determine if they had Internet access.
- **Outside consultants.**
  Firms offer industry-scanning services. Because they are not giving advice and direction, their liability is limited. But be careful. They may sell the same information to your competitors. Hire individuals as opposed to large firms; individuals provide accountability and access to a senior person with contacts. Carefully determine if consultants have carried out e-commerce

and E-Business data collection and analysis before. Determine if they have gone beyond just scanning web sites.

METHODS FOR E-BUSINESS

STEP 1: IDENTIFY THE INFORMATION TO BE COLLECTED

Think of data collection as resembling a sieve or funnel. Start by collecting a wide variety of information related to the business activities. Over time reduce the information that you actually use, while retaining and organizing the data you collect.

What information do you want to collect? Following is a list based on five common groupings. Note that this list is extremely ambitious. It is difficult to obtain some of the detailed information. In this situation, you must be willing to set your sights high and be satisfied with less. If you continue to collect information, you will gradually become more efficient and obtain valuable information.

- Industry segment level
  — Financial health of segment
  — Industry segment leaders and why they are in authority
  — Trends in use of automation and technology
  — Trends in the use of e-commerce
  — Ratios and statistics for industry
- Specific company
  — Financial information—revenue, costs, etc.
  — Trends in the company
  — E-Business strategy being followed
  — Leading achievements
  — Leading issues
  — End products and services— features, cost, quality, variation, availability
  — Customers and their perceived strengths
  — Suppliers and their relationships
  — Leaders and their characteristics
  — Organizational structure, change, and status
  — Infrastructure and its change and status
  — What the company perceives as its strengths
  — Specific activity-related thrusts and strategies
- Business activity
  — Activity organization, distribution, centralization, and roles
  — Changes in the business activity

— Degree of integration among business activities
— Automation support
— Use of e-commerce by the business activity
— Technologies that support modern versions of the business activity
— Perceived importance of the activity to the firm and industry segment
- Technology
  — Extent of specific technology in industry use
  — Reported experience with technology
  — E-commerce software and technology in use
  — Issues in implementing and using the technology
  — Integrating the technology into the business activity
  — Age of technology
  — Competitive technologies and vendors
  — Vendor characteristics
- Country
  — Leading companies by industry in country
  — Financial and demographic data and statistics
  — Restrictions on E-Business imposed by the country
  — Specific laws and cultural factors impacting business activities

Examples of Data to Collect

Following are some examples of specific information that you might collect, along with potential sources:

- Market share by area. Total unit sales and dollars (annual reports and industry surveys).
- Profitability. Return on sales, equity, and assets (financial reports).
- Product. Type, style, color/variety, warranties, price, discounts, and quality—yields and errors (magazines on products and marketing).
- Research and development. Costs, cycle time, and cost versus revenue focus (annual reports and magazines directed toward research).
- Automation used by a company. Technical and web magazines using the company as an example. Look for stories on successful e-commerce software implementation.
- Labor. Cost as percentage of sales, work week and overtime, and productivity—unit and revenue, and mix of workers.
- Organization. Head count, layers, and turnover (outside reports with some information in annual reports).
- Reports of e-commerce initiatives. Annual reports are a good source.
- Human resources. Benefits, bonuses, and training (human resource surveys).

- Capital. Depreciation, lease, fixed asset turnover, maintenance, and cost of capital (standard financial reports).
- Service. Complaints, response time, delivery speed, availability, and order entry (industry statistics).
- Manufacturing. E-commerce, outsourcing, EDI, just-in-time (JIT), locations, and automation (industry surveys).
- Sales. Advertising of the web site, signage, promotion, and sales force (some financial data in annual reports and marketing and advertising magazines).
- Information systems. Costs and head count (surveys of industry use).
- Policies. Depreciation, debt, dividend, financial, accounting, and tax (included in annual reports).

STEP 2: IDENTIFY SOURCES OF INFORMATION

In order to search for appropriate firms, expand your horizon beyond local companies. Consider the following:

- End products and customers. Who makes similar products? Who addresses the same type of customers?
- Industry. Who is in the industry (with "industry" interpreted in a broad sense)?
- Location. By regions of the world, who are leading firms with similar activities?
- Suppliers and distributors. What can you learn from suppliers *to* you and customers or distributors *from* you?
- Vendors and suppliers of technology. What successes have been reported on e-commerce and E-Business?

Much has been written about the global environment. Information on foreign firms is widely available today. Contacts are frequent. The tide of information tends to stress the similarities around the world. However, exercise caution when selecting firms to consider in your assessment. The international dimension adds complexity in the following ways:

- The nature of the global corporation is more complex. Structures of coordination, control, and organization differ widely. These differences are based on the environment of the countries in which a firm has a presence; they are also part of the outgrowth of the organizational culture.
- There are many definitions of e-commerce and E-Business. Many firms try to portray themselves as moving or being established in e-commerce.
- Different industries place a different degree of importance on specific business factors. Examples of these factors are flexibility in operations, control

of risk, customer relationships on a global basis, supplier relationships, the extent to which the products are global, and joint resources and operations. Thus, comparing internationally across industries can be difficult.

- The culture, regulations, and business environment of the country are reflected in information collected. These factors have a major effect on the usefulness and applicability of the information.

The web, libraries, and network databases are two examples of passive resources. In the past, libraries were often viewed as being limited in information, but with the advent of the Internet, web, networks, and CD-ROM databases, libraries with such resources can access much more data.

If you start your search at a college or university library, start with industry surveys and annual reports. Here is a partial list of items to consider:

- Company-specific
  — Annual and quarterly reports
  — Form 10-K (includes five years of information)
  — State corporation filings, such as articles of incorporation
  — Claims in reports of e-commerce initiatives
- Government
  — Industrial Outlook, Department of Commerce
  — National Technical Information Service (NTIS)
  — National Institute of Standards and Technology
  — State government records
- Industry-specific
  — Trade groups (e.g., American Petroleum Institute, American Bankers Association)
  — Encyclopedia of Associations
  — World Guide to Trade Organizations
  — Moody's various industry manuals
  — Standard and Poor's Register of Corporations
  — Business Rankings
  — Registers of specific industries (e.g., manufacturing, energy, banking, and retailing)
- Consultants and advisors
  — Nelson's Directory of Investment Research—investment analysts
  — References of consultants and professionals

These resources can be found on the Internet or in the library. For specific industries, contact the relevant federal and state agencies that regulate or oversee them.

Look for periodicals and magazines that are good resources. A list of business-oriented publications can be found in the Appendix. The magazines *Information Week, Datamation,* and *CIO* offer surveys of firms and highlight leading firms in different industries. One article provided a ranking, total employees, total information systems employees, the information systems budget, the percentage involved in E-Business, and the percentage involved in client–server computing for 500 companies.

Begin your search for additional publications with the following sources:

- Web magazines
- CD-ROM periodical indexes and summaries of articles in technical and business areas
- Business Periodicals Index
- Standard Periodical Dictionary
- Willings Press Guide for international sources

Web and CD-ROM sources are particularly useful because they support key word searches and can print out summaries. Also, CD-ROM sources are usually free. Updates appear on a quarterly or annual basis.

After you have exhausted the library resources on your own, consult the reference librarian about specific questions. He or she may have suggestions about sources available at other libraries or different ways to access on-line indexes of libraries. Reference librarians have also probably done many searches on the web and can impart some of their experience.

On-line networks are another source of information. Access references and articles through Internet or commercial on-line services. There are many different on-line databases. It takes quite a bit of time to find information and to become experienced in quick searches, so consider this only after you have exhausted library sources.

You might then move on to international information. Contact United Nations agencies, the Department of Commerce, securities brokerage houses, the International Trade Commission, or your city's Chamber of Commerce.

Before you begin to collect data by telephone, electronic mail, and in person, organize the information you have already collected. Develop a list of specific questions for the active collection stage. Tailor these to specific companies and business activities.

Begin with firms that are not direct competitors; for example, you might start with technology firms that provide E-Business-related hardware, network components, software, and services. Suppose that you have read an article about a success story in which a company used a specific technique or technology that provided benefits. Begin your information search by contacting the technology or consulting firm and see what information you can obtain. Explain that you are in the early stages of an E-Business project and that you are collecting information on business activities. Be sure to reference the article in your discussion. Ask the company's employee if he or she has a more detailed report. These people tend to be friendly, because you are a potential customer.

Once you have obtained sufficient information from these firms, begin researching suppliers and customer firms. Use employees from your company to obtain names of contacts. Again, the people you speak with should be friendly, because they already have an established business relationship with your organization.

The toughest source to tap is the competition. Using the web and magazines you may glean some names of potential contacts. Contact names can also be gathered through employees or through literature. When you call competitors, introduce yourself and state your purpose. Indicate that you wish no proprietary information. Follow up this call with a formal written request. Be sure to have internal management and your legal department staff review the letter. In your letter, focus on business activities. State that you are not interested in confidential product or manufacturing information. Follow up with telephone calls. This approach can also be used for other companies that are not direct competitors or companies that are in the same industry but in another country.

If you are fortunate, you will receive an invitation to visit one or more companies as a guest. At the end of the visit go over your questions and make sure you have the answers you need. Assume that you will not be able to return for some time. Follow up the visit with a thank you letter and an invitation for you host to visit you.

Visits provide not only an opportunity for direct observation but also a chance to meet with a variety of people in different departments. This gives you different points of view about the firm. Use these visits and contacts to establish a rapport with staff at the companies you contact. These contacts will be useful later when you have follow-up questions. Visits also help you get beyond the e-commerce hype.

Try not to jump to any conclusions from the information you have gathered. Some articles promote a particular e-commerce product or technique. On visits you are likely to be shown the most favorable circumstances. If the company has problems or is very protective, your visit may occur when the plant is idle or running with little activity.

STEP 3: ORGANIZE THE INFORMATION

Store and index articles using a simple alphabetical index of items and key words. Do not worry at this time about any duplications or ambiguity. If you find information that cannot be copied, write it on standard note cards. You could enter everything into a computer, but this takes time and is inconvenient if you are in a library without a laptop computer or power outlet. Additionally, data entry detracts from your focus on sources. After all, there are some benefits to this old-fashioned paper-based approach.

Use your computer to organize information once you have your note cards. Select multiple software packages for this task. Use word processing for standard text (as opposed to idea-type software), because eventually your report and findings will be prepared using word processing. Next, put tabular numerical data into a spreadsheet for graphics that you can later embed into the document. Enter information such as key quotes related to the subject, details of the sources, comments on trends, and notes on issues. If you use specific note-oriented software, make sure that you can easily import data into your word processing software.

As you enter data, you will notice holes in the information. Start a list of data you wish to find. Keep this with you at all times. Contact the reference librarian as little as possible during the early stages of the data collection. When you have defined your list of items, you will be able to use the librarian's time most effectively.

STEP 4: DETECT TRENDS AND KEY FACTORS

Use the information to detect trends and key factors. Some of the major ones considered are the following:

- Trends in e-commerce and E-Business use — look for details here
- Integration to link standard and E-Business processes
- Emergence of dominant leaders in an industry segment
- Global firms with a substantial consistent presence in different countries
- Centralized control and distributed responsibility
- Ability of suppliers to exert cost pressure on competitors
- Ability of customers to exert price pressure
- Ability of competitors to press advantage

Information on the competitors' web sites needs to be organized and structured. You can also prepare a table where the rows are the competitors and other firms, and the columns are periods of time. The entries consist of comments as to how the site has changed. Consult this table prior to going out on another round of searching on the web.

STEP 5: DEVELOP COMPARISON TABLES

From the data, build the following comparison tables:

- **Activity group versus competitive/other firms**
  This table displays in the rows the business activities in the group you are considering. The columns are the names of various national, foreign, and

international companies. Make the far left column your firm. The entry in the table is a rating of how you perceive the company is doing with respect to the activity. Support this information and other tables by footnotes. Each footnote gives the source of the rating, a comment, or the basis for the inference.

- **Activity group versus e-commerce-related technologies**
  The activities of the group are the rows. The columns are various hardware, software, and network technologies. The entry is a rating of whether this technology applies to the specific activity.
- **Competition/other firms versus internal infrastructure**
  This table has the companies listed in the rows. The columns can be a mixture of quantitative and descriptive information. For example, consider the following columns:
  — Investment in plant and equipment (rating of 1 is low and rating of 5 is high)
  — Cumulative investment in plant and equipment
  — Degree of centralization (rating of 1 equals highly centralized and rating of 5 equals highly distributed)
  — Industry rating of infrastructure through surveys.
  This table reveals the comparative investment that firms have undertaken. Footnotes could indicate specific projects that you have uncovered in the search.
- **Competition/companies versus information systems**
  The column headings in this table include information systems budgets, a mixture of information systems employees, a number of information systems employees, a ratio of information systems employees to total workforce, a ratio of information systems budget to sales, a ratio of information system budget to information systems employees, and a ratio of information systems budget to total workforce. This table reveals a comparison of information systems investment. If your firm falls at the extremes of the ratios, consider appropriate action.
- **Competition versus industry statistics**
  This is a standard table drawn from industry averages. The columns include the industry measures. The entry is the value of the measure in terms of the specific company.
- **Competition versus organization**
  The columns in this table are characteristics of the organization, such as head count, head count trends, distribution of head count into production and support functions, and turnover.
- **Competitors versus competitor business activities/technologies**
  While this is similar to the table that gave activities versus e-commerce technologies, this one is specific to competitors. This table should identify

the key business activities as columns. The ratings show how the literature and other sources rate the business activities. Footnotes provide comments on sources and explain why they received the rating. For retail industries, consider using e-commerce, point-of-sale, EDI, Quick Response, scanning, dynamic store inventory, and supplier relationships. For insurance industries, consider using e-commerce, EDI, electronic funds transfer, distributed systems to agents, and business functions, such as application processing, servicing, payments, and claims.

You desire to make an impact on managers and staff, and these tables support a faster understanding. The competitive and industry information is intended to show shortcomings in the status quo and to generate enthusiasm about the E-Business. If your firm is rated as a 2 and is the lowest in a row that has values of 4 and 5, people will see the need for change. Simple text may have little impact.

When you present these tables, do not mix the presentation with information on business activities and other activities. Tables tend to have a major impact, so do not dilute that effect with other information. In one situation, management was hostile regarding E-Business. When management saw the tables and heard what was behind the numbers, they were transformed into supporters.

In developing the end products, circulate a draft of the tables and graphs with dummy or partial data and get people's reactions. You want them to become involved and understand what they will see later. When they finally see the end product, they will focus on the data. They may also suggest other graphs and charts. Keep the number of graphs to less than 20 to prevent information overload. Keep a few good charts in reserve to use later, if necessary.

Focus on presentations, as opposed to reports. Reports tend to be dry and go unread. Put the chart on the right-hand page and the text of bulleted highlights on the left side. This side-by-side presentation is effective. When presenting, use two overhead projectors.

The final presentation to management for the industry assessment can employ the following outline:

- Executive summary—key findings and action items to pursue
- Purpose and scope of the assessment
- Comparison tables of the assessment
- Graphs and tables related to industry and competition
- Cost and benefit analysis

Make a list of the key points you have learned from the external industry assessment. From the information you have collected, choose an example or two that you could present to management. How do these examples apply to your business activities?

E-BUSINESS EXAMPLES

RICKER CATALOGS

Ricker was pushed to the wall in terms of the schedule for E-Business implementation. Thus, there was limited time to assess web sites of competing catalog firms. They did perform a limited number of reviews. In retrospect their web site would have been richer and had greater functionality at the start had they done a survey.

MARATHON MANUFACTURING

Being the first firm in their industry segment to go into E-Business, there were no competitive sites. Nevertheless, they were able, through diligent searching, to find web sites of manufacturers in other areas. They found one site that gave lessons learned. This triggered their desire to pursue lessons learned.

ABACUS ENERGY

Abacus really did not need to consider competitors. However, they did want to see how other firms dealt with suppliers. So they contacted some suppliers with whom they had a close and long relationship. These sites along with the information and comments from the suppliers helped in the requirements for their web site.

CRAWFORD BANK

Crawford was well aware of the banking and financial services web sites. That was what had got them worried in the first place. What they did not know and could not know was what new entrants were planning to enter the market and what other competitors planned to do in the future. They decided to assume that the competition would be very creative—they were right.

E-BUSINESS LESSONS LEARNED

- **Measure yourself not by the volume of data you have collected, but by the holes in the information that remain.**
  Use this very demanding standard to evaluate your work.

- **Subscribe to magazines that are free to qualified subscribers on the web.**
  Hundreds of magazines are available to people in various industries. This can turn out to be a steady mine of information. You will have to fill out a qualification form that indicates your potential as a customer to advertisers.
- **Set up a database to record observations on competitors' web sites.**
  It does the project and organization little good if the observations and analysis of other web sites are maintained in an unorganized manner. Include printouts of the web pages, dates, and analysis.

WHAT TO DO NEXT

1. Develop a list of companies or organizations that are leaders in your industry nationally and internationally. Next, develop a list of firms that are innovators and examples for your critical business activities. You can use the literature to identify the firms and then access their annual reports.

   For each firm develop the following for the past three years:

   - Size of the firm in sales
   - Net profits
   - Capital investment
   - Earnings per share
   - Debt
   - Number of employees
   - Sales per employee
   - Expenditures on automation (if available)
   - E-commerce trends

   In the annual report, to what does the company attribute success? How do the company's business objectives match up to those of your organization?

2. Define categories for your data collection. Examples are specific business activities, types of computer systems, and industry type. A notation can have multiple categories or types. See the following item.

3. Prepare note cards with the following information:

   - Source—publication, volume, issue, date, pages
   - Title

- Author
- Type—category
- Quotes or statistics
- Comments

4. Prepare a table of activity groups versus firms. The entry in the table is the rating you give to that firm for that activity.

Chapter 6

Action 4: Set Your Technology Direction for E-Business

INTRODUCTION

When you consider technology for E-Business, there are two components. One is the technology itself. The other is infrastructure. Infrastructure is significant because in many implementations of E-Business you will make changes in the current business activities that require changes in facilities, layout, and other aspects of infrastructure. Ricker Catalogs ignored the office layout. When they implemented E-Business, they found that they had to relocate the customer service and telephone ordering since E-Business activities are integrated. Marathon realized the potential problem and decided to address it. Abacus did not face the problem. Crawford Bank was able to locate its subsidiary in low-cost, low-visibility facilities.

E-Business establishes new relationships with customers and suppliers. Neither of these audiences can be controlled. There are significant impacts on the technology you select for E-Business. First, what you select must be highly reliable and available. Downtime really means lost business. A second factor is that E-Business mandates that the technology support a wide fluctuation of workload. There will be major peaks and troughs. You must select and size the technology to handle the peak loads—far beyond what you would encounter in internal systems that are more predictable. For some sites, transaction volume can rise by a factor of 10 when an attractive promotion is offered. Ricker Catalogs experienced a growth of 500% during one promotion. This is not unusual.

INFRASTRUCTURE

Both standard business and E-Business activities exist within the infrastructure and depend on the infrastructure for their performance—availability, reliability, speed, cost, etc. If the infrastructure is shaky or is deteriorating, the activities can quickly deteriorate. Many people restrict their attention to technology. However, buildings, telephones, facilities, and other aspects of infrastructure often are equally important. In one project in a department, furniture changes and additional chairs were significant. Moving several groups to adjacent locations was also effective.

Infrastructure impacts the functioning and morale of an organization. Even though E-Business is based on automation, it still relies on the organization and people. If an organization is divided geographically, management and coordination are made more difficult. Similarly, specific problems with utilities can affect the organization's opinion about technology and its reliability.

Managers often separate the infrastructure from the business activities. They do not control the infrastructure, so they do not request improvements. Some managers treat infrastructure as overhead—the less spent on it the better. However, when the infrastructure inhibits the business activities, the impact may be severe. Morale may be low if staff members believe that poor working conditions reflect management's assessment of the importance of the business activities. When the smooth performance of the steps of a business activity is inhibited, there is an excuse for poor performance. You may have heard, "I was late with the work because of the problem in finding parking," or "The network was down so we couldn't do our work." Infrastructure in E-Business has a major role. You will have many smaller shipments with E-Business. You will probably have to stock more inventory of some items. The layout of the warehouse will have to be changed to address E-Business. This occurs even if you outsource the shipping and tracking function.

Here are ways that infrastructure impacts business activity performance and effectiveness:

- Where the work is performed. The physical location of organizations performing work impacts the handling of transactions, the resolution of errors, and the ability to track and manage business activities. Location can also impact productivity.
- How the work is performed. Carrying out a business activity may require telephone or communication lines. It may require movement between locations. Old elevators or poor parking can impact the work. The same is true for unreliable telephone or network cabling and lines. Out-of-date equipment and technology can impact a department quickly and on a recurring basis.

Intelligent changes to infrastructure can result in substantial improvements that are cost-effective, since the cost of the infrastructure change can be amortized across an extended period. The employees who perform the work increase their productivity. They also have fewer excuses for poor work. Physical bottlenecks and interfaces can be improved.

Pay attention to business activity impact and support of E-Business when you make infrastructure changes. You can go overboard on infrastructure changes. You can paint, fix up, move, and build to the point of disrupting the work. The business activity will not necessarily be improved. The worse scenario is that the new regular activities are worse due to a lack of attention to infrastructure so that the E-Business and regular business activities are misaligned.

Carry out changes to infrastructure with a strategy in mind. Otherwise, later improvements can undermine earlier improvements. A street system is a good example of this. A city will tear up and repave a street. Then, a week after the repaving, the street is torn up again for water main work. Of course, when the water main work is completed, the street will be patched, but it is never the same as complete repaving. At several companies each time the E-Business was improved, there were impacts on organization and on location and layout.

THE TECHNOLOGY ARCHITECTURE

Pieces of the infrastructure combine into a structure that can be referred to as the "architecture." The architecture can have as much impact on business activities as the individual infrastructure components themselves. Here are sample elements of an architecture:

- External computers and databases
  — External network interfaces
- Security hardware and software, and firewalls
- Host—mainframe/minicomputers/servers
  — Operating system
  — Database management systems
  — Software applications
  — Databases
  — Interface to network
- Wide area network
  — Hardware
  — Communications
  — Software control
  — Network management
  — Interface between networks
- Local area network
  — File server
  — Print/fax servers

    — Database servers
    — Software
    — Network operating system
- Internet service provider
- Credit card network links and authorization
- Internet software
    — Software tools
    — Custom software
    — Databases
    — Interface to workstation

A technology architecture should specify components for each of the areas in the list below. This list can be used in building a description of your architecture—past, present, and future. E-commerce software, hardware, and network components fit into this architecture. Some of the components you have to address are:

- E-catalog software
- Ordering and market basket software
- Customer profiling software
- Cross-sell software
- Customer service software
- Ordering tracking
- Inventory access software
- Order processing and handling
- Accounting software
- Credit card authorization software and linkage
- Credit card processing
- Firewall for security from Internet access
- Hardware servers and system software
- Network management software
- Use of an Internet service provider
- Web development tools

Here are two examples of computer architecture. The first example is based on mainframe computers, while the second example is more modern and includes client–server technology.

Example 1: Traditional Architecture

- Mainframe computer
    — Proprietary operating system
    — Old, legacy software applications
    — Rigid data files
- Network with proprietary interfaces

The problems with this architecture include difficulty in interfacing to host systems and databases, the necessity for workstations to emulate terminals in order to talk with the host computer, and the inability to view and work with host data easily.

Example 2: Distributed Architecture

- External computers
- EDI links
- Host computers
  - — Database management systems
  - — Standard databases
- Wide area networks
  - — Network management
  - — Network operating system
  - — Internet applications
  - — Combined local and host data

In this example, the information on the host system is accessible. The host and Internet software are sufficiently compatible.

In the above examples e-commerce software could be supported on the host system. You would also require additional servers to support proxy servers, firewalls, and links to credit card authorization, for example.

Technology

Technology is only one part of the infrastructure, but it is a central part, so we will give it attention throughout this chapter. The obvious role of technology is associated with computers and communications, but technology encompasses much more as people use other devices to access the Internet. Most firms who want to succeed in E-Business will have to be sensitive to the limitations and features of these devices.

Any technology requires support. E-Business, unfortunately (or fortunately depending on how you view it), typically requires more support. There are several reasons for this. First, the content of the site will be more dynamic in terms of products and services as well as promotions. Second, in implementing E-Business you will usually require the products of different vendors. You will have to maintain and synchronize the interfaces.

Technology may require operating support (it cannot run without help). It may need maintenance. Periodically, it may have to be upgraded (enhanced). The technology may interface and integrate with other technologies. It must be capable of being monitored, measured, and managed. As an example, consider an electronic mail system. It runs on a network. It must interface to the network, the network operating system, hardware, and the users of the software. It may require an interface with other types of software, such as other external mail systems. The network, of which the electronic mail system is now a part, must be monitored and managed. This example shows many interfaces and a degree of integration.

Technology moves through stages from a raw product to a part of an integrated system. The location of the technology in these stages often indicates whether it is complete and sufficiently mature.

Technologies can be classified into four areas. A technology is part of the backbone if you rely on it constantly. Your business activities may find it essential. Your computer network or a manufacturing system is part of the backbone. All of the E-Business components that were listed above are part of the backbone. Technology is considered as being niche technology if it fills a limited, very specific need for some amount of time. A specific desktop publishing system from an obscure firm is an example. A technology is marginal if its impact is limited (hence, so is its value). Premature technologies are those that lack support structure or completeness. The first handheld computers (personal digital assistants) fit into this category.

Financial and operational characteristics of the firm behind the technology are important in technology assessment. So is the question of whether the technology is part of the company's core strategy. Alternatively, is it just a niche product that later may not be enhanced or expanded? Is it the first in a series of products? If so, you may be at risk since the product line may be canceled if sales goals are not met.

Your goals in evaluating infrastructure are the following:

- Assess the current infrastructure supporting the activity group.
- Examine opportunities for new infrastructure and technology using external information.
- Identify the changes that are most appropriate.
- Develop a scenario for a future infrastructure.

Note that, in parallel, you are examining the business activities in detail. If you complete the infrastructure assessment prior to the internal analysis, you can revisit and refine it later. You are obtaining information on infrastructure while examining what competitors are doing (as discussed in the preceding chapter). Organizational issues supporting the business activities will be considered when you examine the activities in detail (see the next chapter).

The scope of this action includes internal technology and infrastructure; potential new technology that may benefit your business activities and is likely to be available; and technology and infrastructure used by competitors, other firms in the same industry segment, and other firms and organizations who are performing the same functions.

MILESTONES

The major results of this step are the following:

- Comparison tables that assess the current technology and infrastructure and the potential of new technology and infrastructure for improvement.
- These comparison tables make a sharp distinction between the current infrastructure and the potential infrastructure.
- A scenario for how new technology and infrastructure would work together to support the business activities better than the current technology and infrastructure do.

This gives management and staff a picture of how the entire puzzle of technology and infrastructure can be assembled to help the business activities.

METHODS FOR E-BUSINESS

STEP 1: COLLECT INFORMATION ON THE CURRENT INFRASTRUCTURE

Begin by considering the current infrastructure. Here is a comprehensive list. Not all of these will apply to your situation.

- Physical location of all groups doing the work (important for E-Business since disparate groups may have to work together)
- Warehouse locations
- Warehouse layouts
- Condition of their workplace in terms of space and specific location in a building
- How the work proceeds physically through the space (impacts the efficiency of the business activities which must be in tune with E-Business)
- The proximity of the managers or owners of the business activities to the activities themselves
- The current voice telephone system
- The current technologies in place and their structure and organization
- The condition of supporting equipment
- The support that the infrastructure receives (new skills and people may be required to support E-Business)
- The architecture—how the parts of the infrastructure come together

While this step may seem simple, it is valuable for the following reasons:

- The business activities depend on the infrastructure.
- Infrastructure, as compared to activity steps, politics, and organization, is an

area where money talks. That is, improvement can result from the expenditure of money.

The internal assessment provides a list of issues to be addressed and identifies the major infrastructure components relevant to the business activities. You can also match up the changes in infrastructure to the E-Business implementation.

The infrastructure assessment is based on internal and external information. Pursue gathering information in both areas in parallel. To begin, make a list of infrastructure items that apply to the business activities today. In a typical organization, much of the technical infrastructure should exist in documents. Figure 6.1 lists potential infrastructure components. For each relevant item you will want to know the following information:

- The current state in terms of operation
- The performance of the item
- Existing problems and issues that are known but have not been addressed
- Recent actions taken to fix or improve the situation

This is a view of the infrastructure itself. Also look at what supports the infrastructure. Here you might consider some or all of the following:

- What is the recent history of work on infrastructure?
- What has been the extent of redoing work or failure to pass inspections and tests?
- How are priorities set for work assignments? Are priorities based on business activity or organization benefit?
- What is the method for requesting, approving, monitoring, and evaluating work?
- How has the level of work changed in volume and nature in the past two years?
- What outside management and support contracts are in place?

Gather this information by checking documents on file and, more importantly, by direct observation. When you are reviewing the business activities (as discussed in the next chapter), observe how the staff and business activities interact and depend on the infrastructure.

A two-step approach is recommended. First, observe the work passively. Second, interview people involved in the work. During interviews, ask what people would do about the infrastructure. People often take it for granted and have no ideas. Make suggestions, such as fixing up facilities or modernization. This will likely lead them to give more information.

Mainframe/minicomputer/servers
    Hardware type
    Operating system
    On-line systems monitor and control software
Network components
    Network utilities
    Network management software
Software
    Database management systems
    Fourth generation languages
    Languages and compilers
    Software development tools
    Software configuration management tools
    Security software
    E-commerce software
Wide area network
    Internet service provider
    Transmission/cabling
    Network operating system
    Network utilities
    Communication protocols
    External communication protocols
    Network management software tools
    Network hardware (hubs, routers, bridges, etc.)
    Network software
    Testing and monitoring equipment and software
Local area networks
    File server hardware
    Network operating system
    Network cabling
    Network protocols
    Network utilities
    Database server
    Database server operating system
    Network interface cards (NICs)

Figure 6.1   Examples of Technical Infrastructure Components

STEP 2: DETERMINE OPPORTUNITIES REGARDING INFRASTRUCTURE

Identify opportunities to improve infrastructure by looking outside the organization. Sources of information include those identified in the previous chapter (the web, competitors, industry surveys, literature). To these add visits to trade shows and contact with vendors. You also want to find out how firms changed facilities with E-Business implementation. Opportunities range from new products and services to systems and technology, and new methods for supporting business activities.

Assemble articles and go through them. Include early articles so that you can see trends. You seek to understand how a technology works from a high-level perspective. Gather information on the advantages and disadvantages of a technology, as well as what competitors exist. Determine the state of the technology, whether it has potential for your organization, and the extent of its applicability.

In the previous chapter you covered the steps in understanding the activities in the industry and how competitors carry out work similar to yours. You not only can gain knowledge of what they are doing, but to what effect. You would like to have their lessons learned from a specific technology or infrastructure idea. Of course, this is very difficult to obtain because it represents proprietary information.

Consider approaching contact vendors who supplied the technology to the competition. Ask them for contacts for lessons learned in implementation and operation. This information is invaluable in reducing your learning curve and avoiding blind alleys.

In the past, the roles of the suppliers of the technology and infrastructure were limited to providing information and supporting their products through installation and operation. With today's more complex technology, this limited role is not practical. You rely on suppliers and vendors for technical information and support. To an extent, you also rely on their advice related to the appropriateness of their products to your situation.

Here are some of the tasks suppliers could perform to assist you in determining infrastructure opportunities:

- Provide relevant technical information about the product. Provide the supplier with sufficient information to judge which product version and information best fits your needs.
- Provide references, lessons learned, guidelines, and other management type of information about the product.
- Participate to a limited extent in explaining how you could use their product in your work. This requires a general understanding of the business activities and infrastructure.
- Supply consulting support in determining benefits, issues in implementation, and determination of exact features required.

These roles can only be effective with information sharing among suppliers and the organization. There is some risk that a supplier might use this information later, but in most situations the risk is minimal. After all, you will be changing the business activities for E-Business.

STEP 3: DETERMINE HOW INFRASTRUCTURE CAN SUPPORT E-BUSINESS

Infrastructure support can take several forms. First, the infrastructure can provide the environment within which the staff performs the work. Buildings and furniture are examples. E-Business supports greater integration of functional organizations. This has an impact on facilities. Second, the infrastructure can provide the means of performing the work. A simple example is a telephone system. For E-Business you will want the staff to have access to the company's web site. Each of these increases the role of the infrastructure in the performance of the business activity.

Try asking about what the ideal infrastructure would be. The answer you come up with will point to the ideal environment for the new business activities. Start with the E-Business strategy that you selected and then determine the impact on the organization. From the organizational impact you can then judge the facilities' requirements and consider an improved infrastructure.

STEP 4: NARROW THE RANGE OF ALTERNATIVE TECHNOLOGIES

To see the potential of a specific technology, assume that the maximum effort and options of the technology are available and implemented within the business activities. Develop a technology assessment of the maximum that technology could do for the business activities.

Avoid technologies for which people emphasize standards. This may mean that there are no agreed-upon standards. The lack of standards or warring groups advocating different standards point to incompatibilities between different versions of the products. Two examples are EDI and the operating system UNIX. Both technologies suffered from a lack of standards. These comments apply to web development tools as well.

Discard technology ideas that have some of the following characteristics:

- Only a few companies use the technology. Each candidate technology has a target audience. If a substantial number of companies are avoiding a technology ask why. Besides the potential existence of problems, this technology will give you few lessons learned.
- The technology or infrastructure will require too many resources for implementation.
- The new technology or infrastructure is counter to part of your architecture and standards that do work.

- The learning curve is too high and the time to gain proficiency is too long to be feasible.
- Integration with existing technologies and infrastructure is problematic.

Also beware of a new technology that is generating much excitement. Excitement is in no way equivalent to results. Look further than surface enthusiasm to see whether the technology would be effective in your company. In E-Business there are many new products. Often, it is the lack of integration that kills off the technology.

Group the technologies or infrastructure ideas that now remain. These groups will be useful because it is likely that you will want to select and implement more than one change. The groups will allow you to implement a set of changes over time that will improve the transactions.

There are several ways to group technologies, including the following:

- Those that use the same area of infrastructure
- Those that will be supported by the same organization
- Those that impact the same business activities or their steps
- Those that overlap with the same parts of the architecture

STEP 5: DEFINE THE INFRASTRUCTURE SCENARIO

To organize and structure all of the information gathered, develop a scenario. An infrastructure scenario should cover components, architecture, and support structure. Specify all of the major components of the candidate infrastructure. Include elements of the current infrastructure that remain and new elements in a candidate infrastructure. Define how these components go together and how they relate to each other. This is the architecture. For E-Business you first have the existing architecture and the list of e-commerce components. Your challenge is to fit these together.

The architecture must be supported. Specify the composition and organization of the support structure for the infrastructure and architecture.

To gain practice and useful information, develop a scenario for the current infrastructure. This will provide a useful means for comparing the effects of potential changes. Keep in mind that changing infrastructure takes time. Such changes require more than just planning, getting approval, and doing the work. Permits and permissions as well as legal agreements may be needed. These items tend to stretch out projects. Be realistic in your scenario about the extent of change that is practical in a limited amount of time.

Here are some specific activities for E-Business architecture:

- Define the overall network, both internal and external. This gives you the architecture fabric.

- Place the e-commerce hardware and system software in the network.
- Determine the placement of the e-commerce software in the network.

STEP 6: MAP THE SCENARIO AGAINST THE CURRENT BUSINESS ACTIVITIES

Mapping the scenario against the current business activities will indicate relevance of the changes to the current activities. (At this point, it does not indicate the true benefit. That will come later after you have developed a definition of the new activities.) Note that you want to have an idea of the available technology and infrastructure for E-Business. You will be embedding the technology and infrastructure in the new business activities. Here are two ways of mapping or developing the relationship:

- Activity steps versus architecture. Specify what elements of the architecture support specific activity steps.
- Issues and problems with the current business activities that are infrastructure-related. Will changes to the infrastructure by themselves alleviate some of the issues faced by the organization, relative to the current business activities?

STEP 7: DEVELOP COMPARISON TABLES

These comparison tables will help in evaluating, selecting, and getting consensus on which changes to infrastructure and technology are most appropriate to your organization. The tables are based on the following variables:

- Activity group. The group was identified in Action 2. This is the list of individual activities.
- Current infrastructure. The elements are the key infrastructure components that have been identified as issues during the assessment.
- Candidate infrastructure. Also developed in this chapter, this is the scenario of the new infrastructure.
- Technology. This is the list of either current or candidate technologies.
- Architecture. These are the components that hold the infrastructure and technology together.
- Degree of automation. This is subjective; it identifies the extent to which individual steps in a business activity are automated.

These factors now can be employed to create a new series of tables. Table entries are on a scale of 1 to 5, where 1 is low and 5 is high. The tables are as follows:

- Business activities versus candidate infrastructure. The rows are the business activities in the group. The columns are the elements of the candidate infrastructure. Specific columns should include all changes. The entry in the table is based on the degree to which the infrastructure element supports the specific business activity. This table indicates the degree of fit.
- Current infrastructure versus candidate infrastructure. Rows and columns must contain corresponding elements. The entry in the diagonal of the table indicates the degree of change or improvement. The nondiagonal elements indicate the extent of compatibility (or difficulty of interface). This table indicates the degree of technology or infrastructure fit.
- Candidate infrastructure versus technology. The rows are the elements of the candidate infrastructure, and the columns are the new and old technology to be used. The table entry is the extent to which the infrastructure depends on the specific technology. What fits?
- Organization versus infrastructure. This can be two tables—one for the current and one for the future infrastructure. The rows are the organization groups and the columns are elements of infrastructure. The entry is the extent to which the organization uses or will use the infrastructure. Identifying who uses what is valuable in marketing to see who will benefit from infrastructure changes.
- Candidate infrastructure versus architecture. The rows are the elements of the new infrastructure. The columns are the main components of the architecture. The entry is the degree of fit between the architecture and the infrastructure.
- Architecture versus technology. This is an important technical table. The architecture is the structure of the technology. The key elements of architecture (e.g., wide area network, local area network) appear as rows. The columns are the technology identified in the assessment. The entry in the table is degree to which the technology is critical to the architecture. (Or, which technologies are you dependent upon for success?)

Note that you can expand to additional tables as you consider the current business activities and formulate the new activities. Whether you construct these tables is not important. What is important is that you compare the entities defined in the tables, because infrastructure changes cost money. Prepare a compelling argument so that you can obtain approval.

Keep in mind that you have been working on three interrelated activities—competitive and industry assessment, infrastructure assessment, and evaluating the current business activities and organization. In each area, the goal is not only understanding what issues and problems need to be addressed, but also what changes are possible and what benefits might accrue.

The presentation to management will focus on the potential for infrastructure and technology improvement. You want to get management excited about this potential. You also want management to understand the limitations with some technology that is premature or that has not proven itself.

Use the tables and scenario described in this step. You can add a list of benefits claimed from the infrastructure and technology which is based on what others have cited. You can also add a rough estimate of costs, though you do not have sufficient information to supply detailed costs. Give costs in terms of the order of magnitude.

E-BUSINESS EXAMPLES

RICKER CATALOGS

Ricker had a relatively old internal system for order processing. They decided to use the same basic software for customer ordering on the web and for traditional telephone ordering. Their customer service system was an older system that linked to the same database. It would have to be modified to link to the database of the new system. Ricker underestimated the extent of hardware to support E-Business. Fortunately, they could acquire this hardware later and install additional servers.

MARATHON MANUFACTURING

Marathon thoroughly planned for the hardware and the warehousing. They did not take into account that the smaller firm customers would order a much different mix of goods. This caused problems and adjustments in the warehouses.

ABACUS ENERGY

Abacus faced no real physical infrastructure problems. The web application was custom built so that the major problems and issues occurred in the interface to the old purchasing system. The problems grew so bad that Abacus had to later begin the replacement of the purchasing system.

CRAWFORD BANK

The bank had existing systems for lending. There was an application system, a servicing system, a collection system, and a loan accounting system. The web system was built from the ground up. As with Abacus the major problem was in the interface to the older systems. While there was pressure to replace the old loan accounting system, this was deferred until they found a new software package that supported a wider range of lending and leasing products.

E-BUSINESS LESSONS LEARNED

- **Be wary of just-announced technology.**
  Technology requires a support structure, which takes time to build. Seasoning is often necessary to work out its problems and errors. Time also allows users to gain experience. If a technology is too new or is complex, articles are necessary to explain what the technology is and how it differs from other products. Be cautious about investing in a technology that is featured in articles. Many of these articles are placed by the developers of the product.

New technology finds its natural level through a combination of marketing and customer acceptance. As a technology emerges, the attention moves from definition and standards to guidelines on how best to use it. Organizations receiving benefits will start discussing the technology application and benefits.

- **You do not need a deep understanding of the candidate technology.**
  The time and effort required to deal with the detail diverts you from important issues. Being involved in detail places you at the wrong level. You begin to think too technically about the technology.
- **Look for structure behind the technology.**
  Behind almost all technologies there is a structure. This structure is composed of the technology components, service, and support. The quality of a technology is directly related to the quality of the structure.
- **The largest hidden costs of infrastructure are maintenance and support.**
  Parts of the infrastructure become obsolete; other parts deteriorate with use and time.

WHAT TO DO NEXT

1. List the top five infrastructure improvements that would benefit the business activities. This will help you focus on which changes have the greatest potential.

2. You have identified several candidate technologies. First determine how these technologies relate to each other. In the table below, write the technologies in the

rows and columns. Since this table is symmetric, you only need to fill out half of the table. For each entry use the following codes: N, no connection; M, must have both technologies (row and column); B, beneficial to have both technologies, but not essential; and C, conflict between the technologies so that you must choose.

Technology

Technology				

3. What are the potential benefits of the technology to the activity group you have selected? Construct a table in which the business activities in the group are rows and the most promising technologies are columns. Use a 1 to 5 scale (1 is no benefit; 5 is essential). You will use this table as part of the effort to show people how important the technologies are.

Technology

Business Activity				

4. How do the technologies relate to the infrastructure and technology you already have? This is important because it will show you what you have ahead in replacement, conversion, and interfaces. List the technologies as rows, as before. List the key elements of the infrastructure as columns. Enter the following codes in the table: N, no relation; R, the technology replaces the current technology; I, interfaces are needed between the technology and this element of the infrastructure; and C, conversion of data is necessary for the new technology. Note that multiple codes are often needed. If you do not know what to put in the table, you have just found areas where further data collection is needed.

Infrastructure

Technology				

5. Which parts of the current infrastructure are causing the most problems? Construct a table of business activities in the group (rows) and the infrastructure (columns). The entry will be either N (not applicable) or 1 to 5 (1 means the infrastructure is beneficial for the activity; 5 means the infrastructure is a major problem for the business activity).

Infrastructure

Business Activity				

6. Assume that you have selected the best combination of technologies and that you are combining them with the current architecture to form a new architecture. It is useful to be able to explain the benefits of the new architecture for the business activities. Construct a table in which the rows are the business activities and the columns are elements of the architecture or new infrastructure. Place a number from 1 to 5 in each cell showing the benefit of the architecture element to the business activity (1 is low or no benefit; 5 is great benefit).

Architecture

Business Activity				

Chapter 7

Action 5: Collect Information for E-Business

INTRODUCTION

Critical success factors in E-Business implementation are collecting data and interacting with people through interviewing and observing the work. While these factors are important in traditional projects, their stature increases in E-Business because of the following:

- Implementing E-Business tends to be political. There is fear among the people involved in the work as well as in middle management and supervision levels. What you say and how you work can have a direct impact on whether the project will succeed.
- Collecting data is the principal means of interacting with the staff and management involved in the business activity. Keep in mind that you are not only collecting data, but also showing your confident attitude. During this phase of E-Business implementation you create an impression with the staff who will make your changes succeed or fail. Through the interaction you have the opportunity to get people involved and committed to E-Business.

While traditional methods of interviewing and data collection have value, a total approach that supports E-Business politically is more effective. Change and implementation are marketed at the same time that information is collected. You are also doing informal training as to what E-Business is. Note that with pressure to implement E-Business, the detailed data collection effort discussed here is conducted in parallel with the infrastructure assessment and the competitive and industry assessment.

Here are some ways to gather information:

- Observe the work or organization. This provides direct raw data and allows you the opportunity to ask questions as you observe. In certain business activities it can be misleading depending on factors such as the time of day.
- Interview people. This is the traditional approach. However, interviewing has its drawbacks. First, people can only tell you about things that they can remember. Second, their memory may be faulty. Third, people respond subjectively. Fourth, if several people are being interviewed at once, some may not speak up.
- Use passive sources. Discussed in Action 4, this is the search for information through libraries, the Internet, and archives. Also, here locate and review any training materials and procedures related to the business activities.

Given the significance of data collection, carefully plan, carry out, and follow up after each contact. Be up-to-date on the project at all times so that you can respond to questions about the project. Consciously learn and take information in and continually add to your cumulative base of knowledge.

On the surface this seems straightforward. You can read through the files, observe the work, and conduct interviews. Usually, there will have been previous studies. However, even with the blessings of upper management, you are likely to encounter the following problems:

- **A lack of understanding how E-Business will change people's work.**
  They may think of what you are doing is only narrow improvement—within their department. Since you are focusing on cross-department activities in general with E-Business, keep putting the attention on this and not on individual department characteristics.
- **Previous efforts have been tried at improvement, but no action was taken. How will your effort be different?**
  Prior to conducting interviews, you will learn about actions and items that were not implemented. It is important to discover the attempts, discards, and failures. These can be as important as knowing what worked. Be different. Do not promise them results. All you can honestly say is that you are going to make your best effort, and that you will need their support.
- **You are so dependent upon the existing technology and the existing systems that it would take forever to implement E-Business.**
  This is a rational excuse. The current business activities may depend on old systems. These systems cannot be changed easily, so it follows that the activity cannot be altered. Respond by expanding the scope to include changes in technology. Try to get them to think on a larger scale. Point out that in some cases, E-Business systems are established in parallel to existing ones.

- **The managers see no need to change the status quo, including shadow systems.**
  First, do not directly attack how they have done their work. They have a lot of their ego invested in the current business activities. They even may be very proud of their shadow systems. Gather as much information at the bottom as you can and pay less attention to the middle. Middle management is often remote from the real action and transactions.

It is useful to get these issues out on the table at the start. You must be able to deal with them at the beginning of the activity. Whether you employ these or other methods of dealing with these problems, you will need to be overcoming these barriers along the way.

Your purpose goes beyond gathering data. You seek the best approach for E-Business implementation. You also want to build up a relationship of trust between the staff and yourself. Successful implementation rests upon the trust created during data collection. Of all of the activities in the E-Business project, this is the one that you should do yourself. You can involve other team members, but do not delegate all of the data collection.

Another purpose in data collection is to try to get the people to see how the combination of old systems, workarounds, exceptions, and shadow systems is not good for them. It may barely work, but it hampers their creativity and interest in the job. It is this action where you must psychologically pave the way for change. This is most easily done at the bottom where the work is performed.

The scope of the work includes the following:

- Files. These are not just work and project files, but also files associated with the business activity, such as customer or order files. Additional files pertain to infrastructure and organization. Look for copies of complaints from customers or suppliers.
- Forms and procedures. Consider both informal and formal (official) forms. An informal form might include copies of checklists.
- The business activities themselves. Direct observation and even participation in the business activities are intended here.
- Infrastructure. You can gain information through interviews, files, and direct observation.
- Organizations. You want to gather information about the organizations involved in the work. Political, personnel, and structural information can aid in understanding problems and issues. This is true even if you do not intend to change the organization.
- Customers and suppliers. In certain circumstances when the business activities touch the customers and suppliers, contact them and observe their interaction with the work.

MILESTONES

You want to achieve a detailed understanding of the activity group that you have selected. You want to be able to perform analysis and define the E-Business activities. You definitely do not want to return to collect more data again and again until implementation. Another major end product is mental—you want the people involved to recognize the need for change and streamlining.

Here are some of the end products of a detailed understanding of business activities. You want to keep your E-Business strategy in mind as you collect the data.

- Definition of specific individual steps in the activities. This is the technical core of the work. Level of detail is an issue, as is the handling of exceptions. An exception is a variation of the business activity to address a specific transaction.
- Additional comparison tables that reach down to the detailed steps of the business activities. These tables follow from previous chapters.
- Understanding of how infrastructure and technology either inhibit or aid the work. The support structure is key to change. This also gets at the shadow systems.
- Understanding of the involvement of the organization and staff in the business activities. Attitudes toward the business activities, staff turnover, and supervision are important factors.
- Determination of tactical issues addressed by the new E-Business activities. If you develop a wonderful new E-Business concept that is strategically correct, but it fails to address tactical issues for the staff doing the work and addressing E-Business, you fail.
- Identification of the detailed changes attempted and carried out in the past. Lessons learned are always important.
- Understanding of the politics and self-interest surrounding the business activities. What are people's opinions and feelings? What is important to them?
- Ideas about feasible approaches for implementation. While you do not have the new E-Business activities defined, you can think about marketing and implementation based on your knowledge of the environment.
- Identification of immovable constraints. Where are the real boundaries of the business activity? You may uncover policies that have to be changed to eliminate exceptions.
- Established relationship with the staff involved in the work. Rapport and trust will make the effort easier and lead to successful implementation.
- Fit of the business activities to your E-Business strategy. This document identifies the potential problems in the current activities in terms of their

impact on E-Business implementation. You will expand on this in the next chapter.

Time spent doing the data collection should be limited. Most projects involve less than 20 days of direct data collection. But it may be impossible to gather all of the data and see all of the people in this time. Another form of measurement is elapsed time. Elapsed time is three to four times what the collection effort takes, if you do all of the steps suggested in this chapter.

METHODS FOR E-BUSINESS

STEP 1: DEVELOP THE PLAN OF ATTACK

First, develop a plan for the work. This helps you track your work and determine when you are falling behind. Also, the plan forces you to give structure and organization to the work. The plan of attack should include the following:

- List of potential contacts and their profiles. Profiles might include titles, roles in the work, telephone numbers, locations, and any comments about their role. Make sure to include all of the king and queen bees who have been doing the work for many years.
- Log of documents. As you collect information, you will see many documents, reports, and forms. Log these. This not only helps in organization but will help you respond to questions later. You will want these to help you make the case for the new E-Business activities.
- List of issues in the project. These are issues and opportunities that surface during your work. Give a summary title, the date the issue appeared, a short description (if the summary title is not sufficient), and references to documents or interviews. You might indicate identifiers to show importance and to whom the issue is of interest.
- Approach to the work. This is a description of how you are going to go about your work. Include any methods and tools that you will employ. Tools might include the software you will use.
- Task list summary. This is a list of tasks and a schedule for the tasks. Keep this task to 10 items or less. You do not want to spend much time maintaining this. It is a road map that you can use to make the approach tangible.
- End products. Describe each end product along with an outline and discussion of the use of the end product. Notice that the word "report" did not appear. End products in industry today are often presentations prepared with presentation software.

View the plan as a living document. It is more than a checklist of tasks; it includes some of your key lists. Automate as much as possible. For the issues, use a spreadsheet or database to help in tracking.

Step 2: Set Up for Data Collection

Set up a series of data bases or files on a shared network server. This will ensure that the information is accumulated on a timely basis, that it is consistent, and that people share data. You can use any of a variety of software packages—database management systems, file systems, groupware, or spreadsheets. Choose something that the team members feel comfortable with.

On the software, set up the following:

- Issues database. This is a database of the problems and opportunities associated with the business activities. Over time, it will be refined in detail. Use this to ensure that your new business activities will address these issues. Make sure to identify E-Business-related issues.
- Contacts. This is a file of all contacts for the project. It includes positions, time of contacts, who was contacted, and any pertinent comments.
- Idea database for the new business activities. This is a database of ideas that are gathered from data collection. This will serve to generate the alternative scenarios or models for the new business activities later.
- Benefits. Have a file of benefits that can accrue with change in the current business activities as well as potential impacts of E-Business.
- Implementation considerations. This database includes ideas about organization, infrastructure, policies, and technology that impact the implementation of new workflow and procedures. If these are not recorded, they could come back and haunt you later when you are ready for implementation.
- Tables for people on the team to enter comments. These will lead to the comparison tables.

Step 3: Review the Files on the Business Activities

An early element of the project is to review the files on the business activities. Having the information in the files can reduce the data collection and interview time. It provides the information needed to ask better questions. You also may be able to identify the players and their attitudes.

Here is a typical list of files and file contents:

- Organization charts. You really want several versions of the organization charts. First, you would like the current organization for the company. This includes all groups, not just the ones involved with the activities you have selected. Next, you would like an organization chart from one to two years ago in order to see what happened to organizations and who moved up and down in the company. Some of these people may be critical in your effort. You will be using these organization charts as you consider the real organization (discussed below).

- Customer and supplier letters relating to problems. These can be revealing of the work as well as the attitude of the organization toward its customers and suppliers.
- Departmental budgets (past, current, and projected years) for departments involved in the work. Keep the budgets detailed so that you can determine how much effort the organization devotes to activities around the business activities.
- Annual internal and external reports. You want to know what people say publicly about issues, opportunities, and trends. The internal reports are those annual progress reports that divisions and departments might make as they summarize the year. Quarterly reports may be available as well. As with organization charts, get several years' worth.
- Previous studies about work improvement, quality management, and other efforts at change. These studies may be hard to find. If a company spent a lot of money on a study to improve itself and nothing came of it, the company may be embarrassed to have evidence of this available.
- Reports on information systems, technology, and infrastructure. This includes the requests for changes to systems. What changes and enhancements have been made? What was the basis for selecting the technology currently in place?
- Studies and analysis of e-commerce and E-Business. This includes any assessments from current e-commerce activities.
- Procedures and forms related to the activities that you will be analyzing. Also look for information on procedures for using any automated systems that support the work. Ask to see old versions of procedures and forms, if they are available.
- Internal memorandums and notes about the departments involved in the business activities. Include any internal or external audit reports and findings, if these exist.

Where will you find these documents? At each step in your interview and data collection tasks, indicate that you want to save time and not ask obvious questions and so will need to access the files.

When you review a file, first write the document identification in the log. Make copies of important documents after you have reviewed the file. Return the original documents to the file. Inform staff members that you have returned the documents. Review the copies later in your office.

When you read the document, read it first for content. Note the issue or subject that is being addressed. Note the person who wrote it, and the audience. As you review documents, look for references to other documents in the same file.

After reviewing a number of documents in a file, you can construct several document trees which show the documents in time sequence. The length of the tree and how it ends are important. If there is no tree, the document went nowhere. The bottom of the tree is the last event. Was the subject identified at the top of the tree resolved at the bottom? Are there many stand-alone trees? This would indicate a possible lack of communication. The length of a tree can sometimes indicate the degree of interest in the subject.

Look for follow-up reports. Are there memos indicating the type of action to be taken with respect to the reports you have found? Or is the file silent? If there is no follow-up, make a note to ask in interviews what happened.

In reviewing files, you are trying to answer some or all of the following questions:

- What are specific steps in the business activity? Which organization performs what steps?
- What is the condition of the business activity?
- What do people feel are the boundaries of the business activity where information is given or obtained from other departments?
- Who are the people who write about the business activity and what are their attitudes?
- What are the attitudes of organizations and managers toward the business activity?
- What are apparent issues regarding the business activities? How were issues resolved? Why was there not a follow-up?

Sometimes, in collecting data, you will find files that are in poor shape or not in one place. In such cases, ask to see the forms. To review a form, consider the following:

- What is the date of the form? Is there a number? The date and number indicate that it is formal form. If it is undated, make a note to ask how long it has been in use. If the date is old, this may indicate that the business activity with respect to the form is stable; alternatively, it may mean that the form's use is infrequent.
- Look at the layout of the form. Does it appear well organized? Prior to computers, form design was considered an art and viewed as important. Today, it is less so. But fields on the form that group together should be adjacent.
- When you view the business activities, ask to see several completed forms. How well does the form hold up under real use? Are there many handwrit-

ten notes and attachments to the form? This indicates that the form is not serving its purpose.

- Are there multiple copies of the form? Is the routing of these copies clear? The copies can give you a trail for following a transaction in the business activities.
- If there are instructions for filling out the form, try to follow these instructions. Are the instructions complete and clear?

Perform similar functions with logs. In E-Business you will probably want to eliminate most logs. You can also address how the logs came into being. Was it because of problems with surrounding departments? Then the log was created as a defensive measure.

STEP 4: COLLECT INFORMATION ON YOUR CURRENT E-COMMERCE AND WEB ACTIVITIES

In many cases, your firm already has an external web site as well as internal intranet site. Review these to see if you can answer the following questions:

- Is there a strategy for the web site?
- Is the web site updated? Has its structure and appearance changed dramatically? If so, why?
- Has there been any impact on the current internal business activities due to the web site (such as more customer calls) and vice versa?
- What is the volume of traffic on the web site?
- What features and capabilities does the web site have? What can customers do on the site?
- What is the extent of web content? How is its accuracy verified?
- Is there an effort made to update the content of the site?
- What are the plans for the site in terms of change?
- For the internal site, are most employees familiar with it?
- How much is the internal site used?

For the external web site, you are doing an evaluation of its effectiveness and the extent to which the organization is committed to maintaining and updating the site. Answering the internal questions will help you determine the extent of ease of use and familiarity the employees have with the web. Look carefully at the second question. All of the way through the remaining steps here and in the next chapter you will be trying to determine the cross-impact between the web site and internal activities. Also, note that the current web site may continue its existence in parallel with E-Business since the existing web site may just provide general company information and not support E-Business transactions. Examples of this occur across the airline industry where you can access the e-transactions from the general company web site. In almost all cases, the general web site came first and the e-transactions were added later.

STEP 5: OBSERVE THE WORK

Experience shows that it is important to observe the actual work. This is true regardless of where the work is being performed. Review the documentation that you have on the business activities and the list of issues. Then contact the supervisor of the work either directly, if appropriate, or indirectly through a manager, and arrange for a walkthrough.

Start the walkthrough at the beginning. Many supervisors often have to give tours. You can differentiate yourself from the more casual observer by asking detailed questions as you go. For example, suppose that you are shown raw documents being received. Ask about the quality of the documents and forms. Ask for samples. Try to get permission to copy a sample of each type. Make detailed notes as you go.

Be aware that the supervisor is granting you a favor. By showing you around in the presence of staff, a supervisor appears to be endorsing what you are doing. A tour can take several hours. The supervisor is under pressure to return to work. Be sensitive to these factors.

Here are some situations you may encounter on the tour, along with suggestions on what to do:

- For some reason, part of the work is not being performed. This is a perfect opportunity to arrange for a return visit.
- The business activity is too big and the tour is getting too long. Cut the tour off at a logical point and arrange to come back when it is convenient.
- The supervisor cannot answer your question. Respond that you can find out later.
- The activity is not in good shape and the supervisor is embarrassed about it. Show understanding and express that it is not the fault of the supervisor or the employees. Assure the supervisor that you understand that all involved are trying to make do with a difficult situation.
- It is obvious that you are being shown only the best part of the business activity. Do not try to get into other areas. Wait for another time to see more.

The meeting with the supervisor should follow the rules of interviewing discussed previously. Point out to the supervisor that if possible you would like to perform some of the work. This means that you will have to be trained to do the work. This is important because it can reveal what formal training and procedures new employees need. You will also learn who has the experience to answer questions and how exceptions are handled.

Another approach is to have them train you as a new employee in their department. This may be less of a burden than you think since they have to do this frequently. This method also tends to put the people at ease.

For E-Business you will be paying particular attention to activities such as ordering, order processing, customer service, order fulfillment, shipping, and returned item processing. Follow the above suggestions. Also, you will want to ask about any impact from the web. Find out if the employees are familiar with the web site. You might detect differences in what customers ask when ordering and during customer service.

Later Work Visits

A basic rule is *do not disrupt the work*. Allow people to do their work. Once you start with recurring visits, plan on going to the work location for a portion of every day. As you keep showing up, people will accept you. If possible, dress like the people who are doing the work.

Adopt the terminology of the organization with which you are working. Demonstrate respect for and affinity with the departments. Adopt some of their key words, phrases, and abbreviations and use them in everyday conversations. Recognize and respect the fact that every organization has its own jargon.

Observe the work in periods of peak and calm. An "average" probably is a myth, but people can tell you what are the peak and slow times of the day, week, and month. Observe several examples of peak and slow times. Observe the workings of the business activity and the interface with the public or other departments.

If you are going to be trained in the work, behave like someone new to the department. When the trainer indicates some experience or lessons learned, show appreciation for this. This will show respect for his or her creativity.

As you perform the work after training, you will want to ask many questions. Save these up rather than posing the questions one at a time. Perform work in 15-minute increments. After each period, write down your notes and questions.

If you cannot be trained because of the physical nature of the work, you will need to proceed through the workflow and observe what people do. When there is no time pressure, you can ask questions and cover topics such as the following:

- How long have you been doing the work?
- How did you learn to do the work?
- What is most unusual type of work you have encountered? This will help you get at exceptions and shadow systems.
- When are there peaks of work? What do you do differently at those times?

- How do you measure how you are doing?
- What happens when there is an error? How is redoing work handled?
- What do you think could be done to make the job easier?
- What are some problems with the current computer system?
- If you could spend money on anything, what would you do?

As you are gathering data, you will be asked about what you are doing. You should indicate that you are planning for the implementation of E-Business that will expand the current web site. Even though E-Business may not expand the site, it will expand the electronic presence of the company. This will be less threatening to employees. Then they will probably ask how E-Business relates to their work. Here you want to indicate that E-Business and normal business must be mutually supportive. This is true even if you will replace the business activities with E-Business because there will be a transition period.

It will take many visits over several weeks to get answers to these questions. Here is an ambitious but realistic list of what you hope to accomplish:

- Identify exception work and how it is handled.
- Identify shadow systems and procedures that employees use to do the work. Gather as much detail on this as possible. It will be essential that your new workflow addresses both exceptions and shadow systems and procedures.
- Draw people out for ideas on improvement and the history of what has happened with the activities over time.
- Identify people who are energetic and eager for change. These individuals can be of great help later during implementation and in marketing the new business activities.
- Identify senior people who know the business rules of the business activity in detail. These people will be invaluable later during the development of new systems and procedures.
- Solicit ideas for improvement and give people credit for their ideas.
- Flesh out the issues that were identified at the start.
- Try to estimate in qualitative terms the benefits of changing the current business activities.

If you establish trust with the supervisor and employees, you may be able to test some parts of the new business activity and procedures on several pieces of work. Prior to each visit, make a list of people and topics for follow-up, based on your notes and observations from previous visits.

What to Do after Direct Observation

First, write a note to the manager of the supervisor to express thanks for the time given you. Then, as with interviewing, make notes after each visit. Update the following:

- Details on the impact and validity of issues
- Documentation of the current business activity in terms of business rules and workflow
- Ideas for improvements to the business activities
- Identification of key employees
- Comments on constraints and conditions you will face when you pursue implementation
- Comments on infrastructure
- Comments on organization
- Comments on technology and systems
- Comments on related policies

As with interviewing, the information should be structured and shared among the project team on a network drive.

After some observations, the team can attend a staff meeting and generate some ideas together. This can occur in a workshop mode. Also, you can show documentation on the current business activity and have people validate the steps. Avoid exotic flowcharts that are not easily understandable.

In reviewing the current work, start to get verbal reaction to issues and potential actions. If you do it generally, it will have little impact and you may arouse defensive instincts. Stick to the detailed level.

After reviewing the business activity, check back with the managers. Indicate that the time spent by their staff was helpful.

STEP 6: PERFORM THE INITIAL ANALYSIS

As you conduct interviews and observe the work, you will be constructing some basic items for analysis:

- Basic information flow for the business activities. You can use a graphics tool. You can also simply use a piece of paper with the following columns: step, description, who performs the step, infrastructure used, forms used, time to perform step, and comments.
- Shadow systems workflow.
- Exception workflow. Exceptions are very important to E-Business since you want to try to eliminate these later.
- Flow of redone work. In E-Business you seek to reduce this to a minimum.
- Issues with a label tagging at which steps an issue is significant. Differentiate between E-Business and standard work issues.
- Rules used to do department work. For Ricker Catalogs this would include guidelines, procedures, and policies for processing and approving orders. At Abacus Energy there would be internal purchasing and contracting guide-

lines. If the business rules are complex or arcane, then you could have more problems in E-Business implementation.

- Input and output volume—peaks, averages, and lows. This will be useful in E-Business because you will get to see the impact after E-Business is implemented.
- Observations on the work itself.
- Steps taken to measure the performance of the business activities and control it. These include both the E-Business and regular versions of the transactions.
- Ideas for minor improvements (e.g., changes in handling of steps, changes in forms).

After some initial analysis, formulate some basic questions about the workflow. Get together an initial list of minor improvements as well. If you can, define a first high-level version of how E-Business would work. Return to the work location and ask additional questions, both to gather additional information and to build up trust and a relationship with the people involved in the tasks.

STEP 7: CONSIDER THE BUSINESS ACTIVITY BOUNDARIES BETWEEN DEPARTMENTS

As you gather information and perform analysis, begin to identify and examine the boundaries of the business activities—where the work begins and ends. You must know about boundaries between departments within the specific business activity as well as boundaries with infrastructure, suppliers, customers, and other departments. Remember that many opportunities for improvement lie on these boundaries.

In E-Business you will be breaking down these barriers. However, you may introduce problems into the electronic part and the modified manual part if you do not figure out how to address the barriers in departments.

Here are some of the problems that might arise at the boundaries:

- The input that starts the work is not acceptable. Errors are already present at the start of work.
- There is no measurement or control at the boundary. There may be no quality assessment.
- The infrastructure poorly supports the work so that the boundary work is flawed.
- Outputs of the work may not be measured or even known. You hear about the transaction because of problems in the next department.
- The surrounding organization must immediately reorganize the work to do anything with it.

For E-Business the inputs, outputs, and controls must be carefully coordinated and meshed. This will then force the streamlining within the nonelectronic version of the business activity. By considering boundaries, you also are making a determination of whether you should follow up in other departments surrounding the business activities.

STEP 8: MOVE TO OTHER ACTIVITIES IN THE GROUP

Ask the same questions and follow the same steps. You want to build a cumulative body of knowledge. Determine whether the issues in one activity carry over to other business activities performed by the same people or using the same infrastructure. For E-Business you will identifying more barriers that will have to be overcome. In addition, you will uncover workarounds and shadow systems for the current business activities. These are not allowed in E-Business. Therefore, you strive to identify them here and then eliminate them later.

Also try to get ideas about which business activities and steps would be the best starting points for implementation. For E-Business this does not assume that the initial ordering and catalog are the starting points. You might want to jump to lessons learned and customer service if these were problem areas.

STEP 9: BUILD COMPARISON TABLES

Earlier comparison tables focused on the business activity level for analysis. This was all that was possible without detailed information. Now you can move down to the individual step in the work. You are attempting to identify which steps in business activity deserve greater attention. Consider no more than 10 steps per activity so that you do not get entangled in too much detail. In the tables where business activity or steps are identified, you should probably build two tables— one for E-Business and one for the new standard activity. For E-Business, steps are most likely functions of the software.

- **Business activity steps versus business activity steps.**
  Earlier you aggregated activities into a group based on common ground.
  Now you will consider similarities at the level of the step. If you can identify similarities, when you define a new business activity you can, perhaps, replicate some of the changes in other activities. You can also validate changes by ensuring that they work across more than one business activity. The table indicates the degree of similarity between steps in the business activities. For the rows, you will often only use the major activity. The columns are the steps of each of the other activities in the group. In the Mara-

thon Manufacturing example, customer service was viewed as key to gain
a competitive edge. Ordering and catalogs were secondary. It was just the
opposite for Ricker Catalogs where the center activity was the catalog and
ordering.

- **Business activity steps versus infrastructure.**
  How does the infrastructure support the work? To answer this you progress
  down to the step level. The rows are the steps; the columns are major elements of infrastructure. Table entries indicate the strength of support on
  a scale of 1 to 5. Alternatively, you could use the table to determine the
  relevance, importance, and shortcomings of the infrastructure with respect
  to the business activity steps. This is an important table since it will begin to
  indicate where infrastructure improvements might make a difference.

- **Business activity steps versus technology.**
  This is similar to the preceding table. However, here you could put the existing technologies in as columns along with potential technologies that you
  identify in Action 4. Relevance and potential can determine the rating for
  the new technologies. Alternately, consider only current technologies and
  indicate the degree to which the technology adequately supports the step.

- **Business activity steps versus issues.**
  In Action 1, you identified a series of business issues relating to business
  objectives. In this table, you are assessing the relevance of the issue to the
  step of the business activity. At first, this appears to be too much detail.
  However, if you find that there is no fit, question your choice of the group
  or the identification of the issues. In place of issues, you could select business objectives. For example, if a business issue reduced paper handling,
  you could rate each step as to the degree to which it is paper intensive.

- **Business activity steps versus organizations.**
  This table indicates which organizations perform what steps in the business
  activity. The table is very useful in showing the ineffectiveness or complexity in business activities that cross multiple organizations. This is also
  important in E-Business because it highlights the interfaces among
  organizations.

- **Business activity steps versus organization pairs.**
  Here you are attempting to identify the interface that a step requires for organization pairs. The rows are the steps and the columns are potential organization pairs. The entry is a description of the interface. This table is critical in your E-Business analysis.

You can modify all of these tables by applying weights to indicate relative
importances or some other weighting factor to produce a new table. Use spreadsheet software, which calculates the new weighted tables based on different crite-

ria. Consider employing weighting factors based on elapsed time of the work in the business activity, importance of the step, and labor effort involved in the step. If you keep coming up with the same general relationships, this sensitivity analysis supports your findings in front of management.

In terms of costs and benefits, you can define how much the current business activity costs in terms of the following, to build a sense of urgency for change:

- Money to perform and support the business activities
- Redoing work and corrections
- Problems in customer service and potential lost business

At this point in the project, management should sense the urgency of moving ahead. With all of the data in hand, they should be even more supportive of the project because they can see what it is costing them in terms of productivity, sales, or efficiency.

Also, at this point you should have identified the interfaces between the critical business activities and the other activities in the group and in the company. You should know if you need to move a activity into or out of the group. You should have a good idea of problems that would prevent the benefits of change from being realized if you improve what you have identified. You will also validate the scope of your E-Business efforts.

E-BUSINESS EXAMPLES

RICKER CATALOGS

For Ricker observation and data collection had to be done in months where business activity was either moderate or slow. Some of the workarounds and exceptions were only triggered in the peak periods. Thus, it was inevitable that some were missed. These surfaced later after implementation and became harder to deal with. The lesson here is to carefully select the timing of this action.

MARATHON MANUFACTURING

At Marathon extensive observation and involvement occurred. Due to the condition of the business activities, the team was overwhelmed by ideas from the employees. While this was politically good, it created more work for the team. The schedule for implementation was affected.

ABACUS ENERGY

During data collection at Abacus, a number of policy issues surfaced that affected the work. This slowed things down. Each policy change and issue had to be reviewed with management.

CRAWFORD BANK

Data collection at Crawford started relatively easy. It turned out to be underestimated. Variations in loan processing and application processing were discovered among the various offices. These had to then be sorted out.

E-BUSINESS LESSONS LEARNED

- **Stress simplification and paper elimination, not job elimination or work change.**
  When people ask what you are coming up with, focus on simplification of work and elimination of paper and handling as you implement E-Business.
- **Keep a sense of humor in considering the work.**
  Too often, people take everything seriously. Ask someone about the oddest transaction or piece of work they have experienced. This cannot only break the ice, but can be useful later in giving examples of workflow.
- **Move between organizations as much as within.**
  If a business activity involves several organizations, make sure that you are visible in all of them. If you do not balance your time between groups, you may lose the support of the departments where you spend little time.
- **Ask people where they spend their time.**
  People will often describe work and their role by importance of the process steps. Thus, they give more attention to review and control steps than collating or entering information. Find out where they spend their time. To improve the work, you must impact the way they spend their time.
- **If people seem to resist involvement, move to keep them informed — avoid isolation.**
  After meeting with a number of department employees, you may sense that some people do not want to talk about the work. They avoid your gaze. Do not leave them isolated. Seek them out and ask them what they think. If they are not talkative, bounce off of them some of the observations and issues that you have identified. Ask them if these are correct.

- **When someone suggests a change, ask yourself why the existing activity exists in its present state.**
  Ask this person to tell you in his or her own words how the change might be implemented. Do not question or challenge. You can learn three things—the new idea, what is important to the person, and language and terminology that may be new to you.

WHAT TO DO NEXT

1. Because many of the business activities cross multiple departments and organizations, it is helpful to construct a table of activity steps versus organizations. In the table, use the following codes: P, performs the step; O, owns or is responsible for the step; and E, involved in the step on an exception basis. Leave the space blank for no involvement. When you define the new business activity (in Action 7), you will be creating the same chart for the new activity with new steps. The two tables can then be compared.

2. Construct a table of exceptions and normal work (rows) versus steps (columns). How an exception differs from normal work and from other exceptions can be seen from the table. Place an "X" if that step is performed for the particular exception.

3. Which technology and parts of the architecture (from Step 4) apply to each step? Construct two tables. The first is for the current architecture and technology. The rows are the elements of the current architecture and the columns are the steps. Then create a similar chart for the new technology and architecture. In the cells within the tables, enter the codes C, critical; B, beneficial, but not critical; and N, not used. These tables show how technology or the lack thereof impedes the current workflow and how new technology can aid the business activity.

4. Relate the business issues defined earlier to the steps to pin down how the business activity contributes to the issue. Note that if an issue cannot be associated with a step, it applies to the overall business activity, policies, or organization.
   Put the issues in the table as rows with general business issues at the top. For columns, enter the steps of the business activity first. Add more columns for policy, organization, and infrastructure. Entries in the table will be 1 to 5 (1 means that the issue does not apply to the step; 5 means that the issue arises directly from the step).

5. Organize the data in terms of the individual steps. Construct a table in which the rows are the steps and the columns are sources of information. In the table are your major findings relative to the steps.

6. Set up an organized file for the information you have collected. Consider using the following tabs for the file:

- Interviews
- Direct observation
- Reports–budget, management, organization charts, workflow
- Minutes of meetings
- File notes
- Other information

Summarize the data according to the following categories:

- Steps
- General business activity
- Current automation
- Shadow systems
- Workarounds

Part III

Define How E-Business Will Work for You

Chapter 8

Action 6: Analyze Information for E-Business

INTRODUCTION

As you collect the information, you will organize and analyze it to define the new E-Business workflow, procedures, and policies. Your audience will include department staff, managers, and members of the E-Business implementation team. In the previous five actions you have collected information that included the following:

- Business objectives and issues
- Details about the business activities
- Details about the infrastructure and organization
- Ideas about potential technology and industrial practice
- Specific problems and opportunities related to the business activities

Here are some steps for a general approach to assessing the information:

- Understand the business activities in detail and from different perspectives. Graphic tools may be helpful for this.
- Model, simulate, or describe the business activity in a way that highlights a specific issue or opportunity. This is important in E-Business when you seek contrasts with standard business work. You want to show the issues that exist and their degree of impact.
- Show the interaction between the activities, the organization, and the infrastructure. After all, many issues and improvements will occur in these areas. Showing only the activity sheds little light on organization.

- Determine the relationship between E-Business and current business activities. While the E-Business activities have not been defined, you know the characteristics of E-Business.

Following is a short list of goals that you want to achieve when you analyze the data:

- Understand how the current business activities work
- Decide on the issues to be addressed through the new E-Business and modified standard business activities
- Structure information to serve as the basis for developing the new activities for E-Business and standard business

MILESTONES

The major end products of analysis are findings that support implementing the E-Business activities. More specifically, you will:

- Understand the current activities, their issues, and their problems in detail
- Be aware of problems that will occur if the current business activities are continued without change
- Understand interaction among activities, organization, infrastructure, and business factors (as revealed in comparison tables)
- Develop an understanding of an approach for the activity group

These end products pave the way for E-Business activity definition.

Other people should be involved in helping you with the analysis. When a person gets involved in the analysis, he or she will be more understanding of the shortcomings of the current activities and the need for change. In general, the more people you involve, the greater the understanding of the need for change. They will be more supportive of change. This is a critical factor for success in E-Business implementation. People who have been performing the same work for years tend to like the status quo and do not see why there is a need for change involving E-Business.

You must be careful in the number of people you involve. There is a trade-off. More people mean that you tend to get more support for change. However, the more you involve the greater the coordination effort.

It may be impossible to do a complete job with limited time and resources. One solution is to consider how much team time and effort you want to spend in each of the following analysis activities:

- Organizing the business activity information
- Performing analysis to look for opportunities to eliminate shadow systems and workarounds, and to streamline in general

- Summarizing the activity information
- Documenting the business activity information
- Performing statistical and mathematical analysis, such as simulation of the business activities
- Conducting follow-up interviews to review analysis results and hypotheses
- Documenting the analysis results

METHODS FOR E-BUSINESS

STEP 1: DOCUMENT THE CURRENT WORKFLOW

You will use work from previous steps. Take a piece of paper and turn it sideways (landscape). The organizations that are involved in the work will be the columns. Put these in the order in which they handle a transaction. In the columns, write a general description of the steps that each organization performs.

Recall that "organization" is a general term. It may include the following items:

- Title of person doing the work
- Person doing the work by name
- Organization and group involved
- Performance measures
- Staffing characteristics

Begin with making a list of transactions in the business activity. Include both common and exception transactions. This list typically may consist of 20–30 different transactions and variations. You will now do analysis at the transaction level.

Take separate pieces of paper for each business activity and transaction and write these at the top of the page. Create seven columns on the page and use an abbreviation of the following as column headings:

- Step number in the transaction
- Transaction step description in common language
- Frequency, volume, and time involved in the transaction step
- Who performs the step
- Automation and infrastructure to support the step
- Interfaces to steps in other business activities
- Issues involved in the step

Issues include the existence of shadow systems, workarounds, manual steps, and exception processing. When you indicate links to other activities, indicate the business activity and the nature and frequency of the interface. Both issues and interfaces are significant since E-Business will have to address these.

The infrastructure you should include depends on the specific business activity involved, but here are some items to be considered:

- Online system
- Shadow department system or procedures
- Network and other software (e.g., electronic mail, electronic forms)
- Computer reports
- Manual forms and logs
- Procedures and policies
- Files
- Location and buildings
- Mail facilities
- Telephones
- Office layout and department staff location

Do not try to include all possible infrastructure items. Only include those that have an impact on the business activity. A test you can employ is to ask the question, "Does this item merely support the performance of the step or does it impact the step?" The frequency, volume, and time involved in each step are important if you want to estimate the cost of performing the work and assess possible improvements.

With the detail you have at the transaction step level, you can move up to the business activity level. Create a new table in which the rows are transactions and the columns are the following:

- Transaction
- Issues (rolled up from the steps)
- Interfaces (summarized from the steps)
- Departments involved in the transaction
- Automation supporting the transaction
- Volume of transactions
- Frequency of the transaction
- Trigger that initiates the transaction
- Indicator of whether the transaction is normal or an exception
- Quality of procedures, policies, etc.
- Role of the transaction in E-Business

The simple column approach has several advantages. First, you can choose to look at the business activity in general or in detail. Second, you can aggregate the detail and move up to the activity level. A simple chart with columns and text fields is all that you need now. If you want to create more exotic charts later, you will have the data. Finally, this will provide detail to both define the new business activity and compare it to the current activity.

When you begin to review your work, first check readability of the tables. Eliminate jargon or common language. See if the details in each business activity cover its scope. Review your notes and make sure that the issues people raised appear in your tables. Look down the organization and infrastructure columns and see if they are complete.

Verify the steps with any written procedures. Attach the blank or completed forms and logs to the appropriate table in which they are referenced. If some of the items in the form or log are selectively entered, place the step number on the form or log.

If you run across a particular business activity that is performed by only one organization, divide the activity into logical segments based on control, different groups in the same organization, or a similar modular structure. Then proceed through the same steps. If the business activity spans several groups, then create a separate form for the steps performed by each department. Make sure to include detailed steps on the handoff to and from business activities. These handoffs will later have to be eliminated or streamlined.

STEP 2: REFLECT THE ISSUES IN THE WORKFLOW

You have identified issues during general data collection. Now that you have identified activity steps, make a table in which the columns are the steps and the issues are rows. Entries to the table will be comments on the issue in terms of impact on the step. Note that these issues are more general than the ones that you identified in Step 1. Consider the impact entries from different views, ranging from the staff involved to managers of groups apart from the immediate organization. "Impact" here means the effect on productivity, accuracy, completeness, control, and other similar factors.

Some of the issues that you uncover may be quickly addressed, prior to E-Business implementation. Take advantage of this opportunity. These are "Quick Hits." Prepare to suggest actions for the short term. As long as they do not undermine the E-Business implementation, intermediate actions both help the business activity and raise your credibility.

STEP 3: VALIDATE THE ANALYSIS

Preceding chapters identified a series of comparison tables. These relate business factors, business activities, organization, and infrastructure and show degrees of relevance, importance, or a similar factor. Use the tables to validate what you have created in your analysis.

To do this validation you may wish to use the comparison table rows, columns, or entries as rows of additional tables and the steps in the business activity or transactions as columns. The entries consist of comments about the column in the row entry.

STEP 4: AUTOMATE THE ANALYSIS

Automating the analysis through spreadsheets or other software tools will accelerate the pace of the work and will enable you to make changes and additions easily without recreating the entire chart. Another reason to automate is political. Use of the appropriate tool, and the inference that you know how to use it, gives you credibility as an analyst. Because you are a user of the tool, you do not have to spend time describing how it works in detail. Thus, you avoid being perceived as being excessively technical.

Roles of Software Tools

Wait until this step to identify software tools because your choice should be influenced by the type of data available and the level of detail needed. Keep your options open until you have collected sufficient information.

Software tools assist in the following:

- Data collection management. As you collect information, you will need to find a way to organize it. Simple microcomputer database management systems are useful. They help you to extract and analyze data.
- Enabling a better understanding. Any software tool that can help you understand a business activity and identify where you should collect more information is useful.
- Analysis and modeling. Software shortens the time taken up by analysis and "what if" analysis. Modeling tools are useful.
- Presentation. Presentation aids are useful for explaining workflow and the business activity.
- Implementation. Software tools aid the implementation of E-Business solutions.

Software tools are particularly important in E-Business because E-Business transactions must be automated to the fullest extent possible. Modeling the transactions will help identify production and business policy bottlenecks.

Trade-offs and Tool Requirements

Tool use involves trade-offs. Before selecting a software application, consider the following questions:

- How long does it take to learn the software?
- How long does it take, and what effort is required, to become proficient?
- How is information entered?
- How are data and results extracted?
- What are the hardware and software requirements?

Be sure that you have the time to learn to use and work with the tools you select. Learning the software tool means that you are able to enter data, manipulate the tool, and achieve results. If the tool requires too much effort in input or output, you will probably not use it much, regardless of its ultimate power.

Categories of Software Tools

Following is a way to categorize software tools that can support your E-Business project:

- **Data management**
  You can use an off-the-shelf microcomputer or network database management system or a fourth-generation language. These tools provide a way to organize data and to export selected data in any of several common formats. They will assist in analyzing and reporting material.
- **Groupware**
  Groupware (e.g., Lotus Notes or Microsoft Exchange) allows you to build files and databases that are available to the project team. The advantage of this software is that it provides functions that are easy to learn. The disadvantage is that, if the software is not currently available in your office, it is unlikely that the E-Business budget can support the implementation of groupware.
- **Analysis**
  Project management. When you think of project management software, you think about defining tasks and milestones to input into the software. However, you can also use the steps in an activity when working with project management software. The software can then be used to check your understanding of the business activity. You can also try out different alternatives by changing tasks, relationships/dependencies, resources, and durations. Use GANTT and PERT charts for analysis.
  Simulation. Workflow simulation software allows you to enter the activity steps and relationships into the software. You can now add durations, resources, and other parameters. The first run should be the current work, which can show its problems and validate your understanding of the business activity. Next, you can modify the workflow and characteristics, and rerun the tool. This step has several attractions: it is quantitative and can

produce interesting graphic results. There is an issue of level of detail, which is addressed later. All simulation tools have mathematical assumptions that can restrict their use in the real world. Additionally, increased time is required to learn this type of tool.

- **Statistics**

  This software is both an analysis and a presentation tool. There are now statistical software packages that interface with spreadsheets and databases. The packages offer a wide range of analysis capabilities which were previously only available on large mainframe computers. They required extensive programming—a situation not unlike the early days of Internet use. Rapid analysis and exotic graphs are pluses, but you must understand statistics and the limitations on the methods in terms of their validity. Thus, this tends to be a tool for the specialist.

- **Presentation**

  General graphics. Standard presentation graphics software will support your presentations. You can paste in graphs and tables from spreadsheets and other software. Become proficient at this because you will use presentation skills frequently.

  Flowcharting and diagramming. This software has free-form and standard drawing templates. There is a range of drawing standards (some of which will be listed). This type of tool requires some specialized learning. If you use such a tool you may need to explain symbols and conventions in a diagram. The tool does not perform analysis. Basically, you enter the data, and it produces tabular and graphic results. Use this category with care as it can be time consuming.

- **Application software solutions**

  Some of the newer application software is sensitive to workflow in a business activity.

- **Process modeling software**

  With process modeling software, the software takes as input the structure of the business activity. It describes each step in the activity in terms of resources needed, time required, and any conditional relationships between steps. For example, if 10% of the items for reaching Step 5 require reworking and are sent back to Step 4, the model reflects this. The output of the model can provide statistics on units produced during a production period of the activity, costs, resource usage, and bottlenecks. You can export output to a graphics or spreadsheet software package. The process models can be modified to see the impact of specific changes to the activity.

  This program appears to be very attractive. You put the information in and let the model do its work. But there are caveats. First, all models involve statistical assumptions on the distribution of arrival of work and the processing time. You may not know these. If you do not have a background in statistics, you may encounter problems. Second, if you use the results in a presentation, you will have to defend the model. Use this tool sparingly, if at all. It is best to use it for your own analysis.

Suggestions on Selecting Tools

When evaluating and selecting software, first identify requirements for tools and determine which tools are available. Then choose implementation software, which will be discussed later in the book. From the standpoint of tool requirements, ask yourself the following questions:

- **Would the tools be most useful in organization, analysis, or presentation?**
  Focus first on the preimplementation tools, although your overall focus might be analysis. When you begin your E-Business project, you will spend a considerable amount of time coordinating people and tasks, obtaining support, and dealing with political issues, which will require a decent presentation tool.
- **Where does the greatest uncertainly lie, in terms of the information gathered? Will any tool help?**
  Tools should help to reduce uncertainty, which may mean collecting more data, so look carefully at the capability to do so.
- **How much time is available to learn the tool?**
  You must become proficient with the tool, because it is dangerous to rely on someone else, especially at a critical point in the project.

Evaluate some highly rated tools in the preferred categories. Mentally simulate how you would use the tool. Ask the following questions:

- What are the data requirements of the tool?
- Is the data available for the tool?
- Do the examples and tutorial cases match your situation?
- Who is using the tool today (perhaps within your organization)?
- What is his or her experience?
- Is technical help available?

Find out if you can obtain the software on an evaluation basis. Set concentrated time aside to work with it. Do not try to enter your information. If you do, you will spend the evaluation period doing data entry. Instead, concentrate on interfaces, parameters of the software, output, and range of capabilities. Use the data that come with the software. Then return the software. If you keep it and do not use it, you will feel guilty and obligated to tinker with it. This effort represents wasted time—time you do not have.

A final thought on selection involves measurement. Devise a method to measure how much you are using each tool, what activities the tool supports, and the effort required in working without the tool.

Justifying Software Tools and Getting Approval

You have some factors working in your favor. First, a body of tools is probably available internally. Second, many software tools are not expensive. Third, many are available on a trial basis, and many packages offer extensive tutorials. However, you still must justify the tool.

Get the support of the team. This involves presenting the range of tools that you will be using to the team. Demonstrate the software and explain its use exactly. Also explain the team's role, if any, with respect to the tool. Explain the measurement of the tool's effectiveness. Solicit suggestions from the team. If one of the team members is an experienced user of the tool, ask him or her to give a testimonial.

To obtain management approval, be prepared to answer the following questions:

- What does the tool do?
- What are the labor and effort without the tool?
- What are the risks associated with the tool, and how you are going to mitigate them?
- How are you going to measure the tool in terms of effectiveness?

Hints on Implementing and Using Tools

Once the tool has been selected, approved, and installed, you will need to implement it. To learn the tool, work with on-line tutorials and examples that come with the software. Copy the example and modify it to fit your method.

Diagramming Standards and Templates

A number of methods have been developed for diagramming and documenting activities. Some of these involve steps; others involve data. It is important not only to decide which software tool you will use, but also to select the appropriate template. A *template* is a graphic representation of a set of rules developed by specific researchers as a standard. Some software packages offer multiple templates. Consider how much time needs to be spent in learning a template before choosing one.

Some popular diagramming methods are as follows:

- *Structure chart.* Using a format similar to an organization chart, each level of the chart breaks down the activity into subactivities and the subactivities into steps.
- *Data flow diagram.* This technique tracks the flow of a transaction in the business activity. The data flow diagram explains each transaction path using specialized symbols. Content labels apply to the arrows connecting the steps and data.
- *Entity-relationship diagram.* This diagram relates data entities, such as customer, product, or invoice. Connecting lines that have specific symbols and labels reveal the relationship between the entities.
- *Warnier-Orr diagram.* At the left of a page is the activity step. To the right is the sequence of items required to perform the step. Farther to the right is the logic to carry out the item.
- *Action diagram.* This appears as pseudo-computer code. At each step of indentation, more logic appears.

The above list is representative, not comprehensive. Some of the items overlap. To be complete, employ three types of diagrams:

- A diagram of the business activity, such as the structured diagram of the activity
- A diagram of the data and relationships
- A diagram that relates the steps to the data entities

There are several benefits of using three types of diagrams. They are established methods and supported by software. The graphs can help you assess your understanding of the business activity. Once information has been entered into a model, changes and regeneration are possible.

The downside of this effort is the labor involved. You must develop very detailed information about the business activity for these diagrams to be complete and useful. If time is short, you may derive more benefit from spending time defining the new business activity instead.

Assuming that you have performed some analysis (e.g., a workflow layout of the business activity and validation of some of the issues in the workflow), you are now ready to receive feedback. Present the results informally. Set the stage for the analysis by first reviewing what you have done. Use the terminology of the departments who will carry out the work. Make a list of jargon, including abbreviations and acronyms, to help others who are not familiar with it.

Start at the bottom of the department and review the workflow step by step. Mark any corrections to the workflow as needed. Return to the department with corrections. People should agree that you have represented the workflow accurately for both normal and exception work. Feedback at this time may give you more information about the work and issues.

Once the workflow steps are finalized, work your way up the organization. This becomes easier as you go along, because you already have feedback from the people at lower levels. Develop a five-minute summary of the workflow that you keep updated as necessary. Use this as the lead-in description of the business activity. To make it more relevant, indicate where issues lie in the flow.

Try to avoid formal presentations. Long, detailed presentations of business activities can be tedious. Your audience may get the impression that you are a technocrat. Provide documentation of the activity in detail in a readable form that your co-workers can understand. Not everyone will read this documentation, but you have given those who want to know more about the project the opportunity to go through it in detail.

You are at the point in the project where the details of the current activities are clear. You also have defined the impact of the technology and organization on the activity. By now, management should be supportive of change but may show some reluctance because the impact on the organization is now starting to become more evident. At this key point, suggest that management start thinking about organization change in the future.

Determine at this point whether you have defined the current activities in sufficient detail. Remember that you will be comparing the E-Business to the current business activity. Have you identified all of the interfaces between the business activities? Are these within the scope of change?

E-BUSINESS EXAMPLES

Ricker Catalogs

Ricker did not engage in extensive formal documentation of the activities. They did work to improve these and clean them up. Rather than documentation, they concentrated on training procedures and materials for staff. They employed project management and spreadsheet software, but not simulation tools.

Marathon Manufacturing

Marathon used process simulation tools and had everyone involved use groupware during the E-Business project. All information was categorized along with lessons learned and information that later would contribute to building the wizard for bidding. The simulation modeling was so successful that it was expanded to non-E-Business analysis.

ABACUS ENERGY

Abacus performed data analysis on each supplier and developed models to determine the workload for the E-Business application. This information was then shared with each supplier.

CRAWFORD BANK

Analysis concentrated in two areas. The first was for E-Business transactions. These were modeled through simulation and statistical analysis. The second was the interface with the bank units. Here each interface was documented. Issues were then associated with each interface. Many of the issues related to communications problems.

E-BUSINESS LESSONS LEARNED

- **Perform actual analysis in departments; do not always go back to your desk.**
  As your analysis proceeds, consider going out to the department and doing the work yourself, if possible. This will put you in closer touch with the people and the work, and will aid your analysis.
- **In deciding whether you have enough data, review the transactions with business staff.**
  This review should be done several times to shake out any additional workarounds and exceptions. Involving the business staff will give you more grassroots support for your effort.
- **Consider the evolution and deterioration of an untouched business activity.**
  In your analysis, consider what might happen over time if nothing is done; for example, experienced people can retire, computer systems become more complex to maintain and more inflexible, more work may be done on an exception basis, or business activity requirements will change. Create a set of tables to indicate the effects of elapsed time on the activity. For example, if you leave customer service untouched, then when the E-Business customers start contacting the customer service staff, there could be problems due to volume.
- **Relate the organization to the business activities in terms of roles and stakes (politics).**
  Roles in the business activity have been discussed, but now you should de-

termine who has a stake in the success of the project. This is important in terms of getting changes approved. Stakeholders should also be part of the feedback regarding the analysis. Are the issues you are addressing important to the stakeholders?

WHAT TO DO NEXT

1. What is the performance of the business activity? Performance can be measured in many ways. The focus here is on activity level performance.

Volume of transactions—total _____

Cost per transaction _____

Types of transactions _____

Assessment of interfaces _____

Assessment of automation _____

Weighted cost of activity by transaction type (apply the total cost to the proportionate volume of transactions)

Type Quantity

_____ _____

_____ _____

_____ _____

_____ _____

Types of exceptions

_____ _____

_____ _____

_____ _____

Volume by exception type

Type Quantity

_____ _____

_____ _____

_____ _____

Weighted cost of normal versus exception work based on volume

Type Quantity

Normal, nonexception _____

_____ _____

_____ _____

Types of errors

_____ _____

_____ _____

Correction effort per error and number of errors by type

Type Quantity Effort

_____ _____ _____

_____ _____ _____

_____ _____ _____

2. Create a table in which the rows will be a list of the activity steps. Between each step insert a step for waiting between steps. The columns will be a list of the normal work and the top three exceptions in terms of volume. Also insert in the table the times that you observe. This table will help you to draw graphs about where the time is spent.

Activity steps vs. work type	Normal	Exception

3. Prepare the following graphs using the data you collected in this section:

- Pie chart on the distribution of errors by type
- Pie chart on breakdown of total cost by transaction type
- Pie chart on distribution of exceptions
- Pie chart on breakdown of total costs of exceptions

You can also use the data you have collected to create the following trend lines on the chart:

- Trend line on total cost per transaction and volume of transactions
- Trend line on number of errors for each type (multiple lines on the chart)

4. Now you can proceed to the activity step level. First, draw the workflow for the normal work and for high-volume exceptions. Place a piece of paper sideways (landscape) and draw a horizontal line to represent the total time that a piece of work or transaction takes to move through the work. Draw vertical lines at the two ends to mark the start and end of the work. Mark the proportional time for each step and for waiting between steps. You now have a graph showing how much time is spent on each transaction. Repeat this approach for exceptions to create a tool for comparing the exceptions and for considering bottlenecks.

Chapter 9

Action 7: Define Your New E-Business Transactions and Workflow

INTRODUCTION

All of the work thus far leads up to defining the new E-Business activities and the changes to the existing activities. It is here that the "rubber meets the road." In Chapter 2 four alternative E-Business strategies were identified. They have been largely left alone until now since the focus was on assessing the current workflow and procedures, and determining the technology approach. Let us comment on how each strategy will affect how E-Business transactions and workflow are defined.

- **E-Business as a separate activity.**
  Here you will be establishing a new business. The business, however, will have most of the same general business activities as the current business. Thus, it makes sense to understand the current activities to define the new ones and to define improvements to the current activities at the same time you are determining the new E-Business ones.
- **E-Business implemented on top of standard business.**
  Following this strategy you will put E-Business in the middle of the current work. This is tricky at best since you will be reusing some of the current business activities for both types of business. As an example, ordering could be based on both e-commerce and telephone call centers. Customer service and order fulfillment may be handled by streamlined versions of the current activities. Ricker Catalogs started this way and then moved toward integrated business activities.

- **E-Business integrated with your regular business.**
  Given time and resources (both big "if's") this is probably the best overall strategy. It makes the activities in both synchronized. It takes more work to implement, support, and manage. This chapter aims at handling this strategy as well as the others. This was the approach of Marathon Manufacturing from the start.
- **Replace some of the existing regular business activities with E-Business.**
  Here you will implement E-Business and more efficient business activities in general. Abacus Energy adopted this strategy with its suppliers. It transitioned much of the purchasing and contracting transactions into electronic form.

The purpose of this action is to define and analyze the new business activities for E-Business as well as improvements to the current work. The scope of the work includes defining alternative new business activities or scenarios, evaluating the alternatives, and comparing the old or current work with the new. Once these steps have been completed, you will be presenting the new business activities to management to gain approval for implementation. The E-Business project has been quite inexpensive up to this action. After this and several planning actions, you will be spending substantial money in infrastructure and technology, and substantial effort in staff time.

In your attempt to develop new ideas, you will be discussing current activities and new ideas with the staff and supervisors involved in the current work, as elaborated in the previous two chapters. You will also cover your new ideas with them to get their input and involvement to ensure that the ideas on the new activities are realistic. This will also help in getting them to provide political support for E-Business. Consider involving specific technical and audit staff to ensure that these areas are covered. In general, review the new business activities with many people several times; this will prevent surprises when you make your presentation to management. More importantly, it will, like the previous analysis, gain political support for implementation.

MILESTONES

The key end products are the new business activities for regular business and E-Business. You will also have supporting comparison tables and evaluations using information on alternative activities. These can be used to compare the old and new transactions, workflow, and procedures, and to see the fit between traditional business and E-Business.

METHODS FOR E-BUSINESS

STEP 1: DEFINE THE DIMENSIONS
OF THE E-BUSINESS ACTIVITIES

Using the analysis you performed, define the dimensions of your business activity. Dimensions are areas of assumptions that are associated with the standard "how," "what," and "why." Different dimensions generate different scenarios. Experience has shown that you can identify the following nine dimensions, or areas of assumptions:

- A. *Activity change.* This dimension is self-explanatory and includes only those changes related to the performance of the work.
- B. *Organization.* This pertains to the organization that owns the activity. It answers the question of "who."
- C. *Infrastructure.* This dimension is similar to the organization dimension in that it pertains to the infrastructure supporting the activity. This answers "what" and "how."
- D. *Resources and importance.* These are related because if an activity is important, management will assign it resources. This pertains to "why" and "what."
- E. *Other organizations.* This dimension includes groups in the same company as well as external firms and pertains to "who."
- F. *Other activities.* This dimension includes any other activities that interface or share resources and support. This dimension answers "what" and "how."
- G. *Management.* This includes all aspects of measurement, planning, and control. It pertains to "what" and "why."
- H. *Automation.* This includes hardware, the network, software, and architecture.
- I. *Policy changes.* This includes changes to policies that impact the activity.

These dimensions are valuable because you need to consider a wide variety of options when generating alternative scenarios. Check yourself by considering alternatives in each of these nine dimensions. Multiple scenarios are often combined and integrated.

Alternatively, a new scenario can be created out of the parts. One way to conduct a visual comparison of the multiple scenarios is to construct a graph that is formed like a star or asterisk. This type of graph is sometimes called a radar or spider chart. Each finger would represent one dimension. The point of each dimension is the degree of change with the particular scenario. You will define an

E-Business scenario first and then the modifications to the current activity. To assess the fit between these, you can graph them using the same spider chart. An example for Marathon Manufacturing is given in Figure 9.1.

When you have done this for one activity in the activity group, move to another similar one. Try to copy the dimensions listed above over to the second activity. Continue in this manner until you have covered each member of the activity group. Next, review all of the scenarios from different views. Consider them from the standpoints of organization, technology, management, and general infrastructure. Determine the issues, problems, and pluses of each activity.

STEP 2: GENERATE SCENARIOS

A *scenario* is a possible specification of a future business activity for either regular business or E-Business. It is a definition of a new activity. A scenario is also called a *model*. Because you can design a new business activity in different ways, multiple scenarios are possible. In information systems, the scenario is the design of the new system and specifies all changes in activities, organization, infrastructure, resources, external organizations, other activities, policies, and control.

To be complete, a scenario should contain the following:

- Description of the steps and how transactions flow through the activity
- The organization and its relationship to the activity

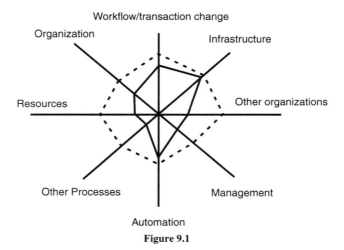

Figure 9.1

- The infrastructure required to support the new E-Business activity under the scenario
- Interfaces with the regular or E-Business counterpart
- Interfaces with other activities and organizations
- Interfaces with customers and suppliers
- How the measurement and management of the business activity will be performed
- Commonality across multiple activities in the activity group

For example, suppose that you were at Crawford Bank and wanted to offer loans, loan servicing, and collections through the Internet. You would start with the current activities. A scenario of the application process for lending could involve the following dimensions:

- Policies. Management specifies more stringent requirements for reporting and measuring loan applications.
- Organization. A single group handles all loan activities. Today, this may be distributed.
- Other organizations. They specify automated interfaces and pass data.
- Automation. Automation requires an on-line system that tracks accounts, interfaces to accounting systems, automatically dials customers, and automatically processes applications.
- How the new system will work. Loan applications will be kept in a single queue. When an employee arrives for work, he or she will log onto the computer. The oldest unprocessed application submitted through the Internet will appear on the screen. The system may have already scored the customer's credit. The staff member has a limited amount of time to view the information and make a decision. The employee enters customer contact actions and results into the system. This is the change for the current process in that even paper applications are handled through the same steps after being scanned or entered.
- Control. The system tracks performance of loan processors and application reviewers.

When developing a scenario, make sure that others can visualize the new activity. They must be able to compare the old and new business activities and see the differences between them from different views: customer/supplier, management, IT, and organization.

Ideas for a scenario might originate in several ways. An interesting and creative way to begin is to consider the impact of drastic actions or triggers (called alternative actions and triggers) on the activity and activity group. These triggers lead to workable scenarios when considered in combination. Other alternatives can result from more traditional methods, such as bottom–up, top–down, and out-side–in models.

Each scenario typically goes through revisions and additions in order to complete and address the entire business activity. Because you are dealing with an activity group, you want to reduce overlap by combining individual scenarios. This is also important in terms of reducing implementation effort.

Using Triggers to Generate Scenarios

A scenario comprises combinations of assumptions. Considering different triggers can lead to scenarios. From a scenario you can derive a concept of how transactions would work in the new activity. Do not attempt to group these actions. Instead, consider them individually. Each of these triggers arises from a combination of the nine dimensions. Each trigger is numbered. You can prepare a table for your triggers by showing them as rows and the nine dimensions as columns. The entry can either be an "X" if the trigger relates to the dimension or a comment. Note that these also overlap in some cases. Also, note that in E-Business you will probably combine several of these.

1. *Eliminate the activity.*
   You would be replacing the current activity with an E-Business one. Before you do this, consider what would be the impact on the organization. What would replace the entire set of transactions? This trigger fits in the management dimension and supports the replacement E-Business strategy. It is an extreme step, but it is valid when applied to activities that are no longer useful.
2. *Place the business activity outside of the organization.*
   This includes outsourcing, which fits into the "other organization" dimension. This would also fit the case where you set up a separate E-Business to do the work. This is a valid solution for generic activities that involve no organization-specific knowledge or competitive advantage.
3. *Starve the current activity.*
   Create the new E-Business activity and let the current one die out due to declining volume. Deny the activity resources for maintenance or enhancement and support E-Business through marketing. Without resources, the old activity will probably be eliminated. This fits within the "resources and importance" dimension category. Politically, this may be the best route for eliminating an activity.
4. *Merge the activity into other ones.*
   This might include dividing the activity and placing its tasks and functions into other ones. This activity fits in the "other activities" dimension. This approach is valuable in an organization that is not competently

handling the activity, but is politically strong. Rather than confront the organization, move the work from under the organization.

5. *Transfer it to another organization.*
   Transfer the intact activity to another organization that is more E-Business friendly. Resources and some elements of infrastructure may accompany the transfer. This supports the separation E-Business strategy.

6. *Throw people at the activity.*
   By putting a large number of people to work on the activity, you can uncover the nonresource bottlenecks in the activity. This demonstrates the benefits of flexibility of a manual system. You might think that this is not viable for E-Business, but it is. You may want to employ more people in customer service in the initial operations of E-Business. This alternative makes sense in E-Business both for implementation and to cover activities that relate to E-Business but will not be automated in the first wave.

7. *Start with management measures and work backward.*
   Management measures include performance measures and statistics reflecting the business activities. This is a variation of the "work backward from outputs" approach. In this case, you work top-down (beginning with management) on the activity. This alternative considers whether the business activity is significant. The activity then feeds the measurement approach.

8. *Throw money at the business activities.*
   Throwing money at an activity often makes problems worse. Money brings new resources into the activity. These resources are then incorporated into the activity. This has what happened for some e-commerce firms when they expanded too fast and then could not handle all of the types of work in a computerized manner. Providing a resource with money is useful, however. It helps you set priorities for what the activity requires. It also reveals problems that cannot be fixed by resources, which include policy problems, organizational issues, and interfaces with customers and suppliers.

9. *Copy the best competitor.*
   This is attractive for E-Business. Use the information you gathered from observing a competitor (see Action 3) and generate a scenario. This trigger reveals goodness-of-fit, but it never seems to fit exactly. Attempt to discover why and you will find some of the underlying factors that differentiate you from your competition. For example, Marathon and Abacus were able to discover basic internal policy problems.

10. *Outsource the activity.*
    When a company outsources, it hires outside vendors to do some or all of the work in an activity. When considering outsourcing an activity, first ask which activities and work are specific to the organization and which

are generic. Examples of organization-specific activities include those that depend on company knowledge, procedures, and sensitive data. The outsourcing trigger is a good indicator of the degree to which the organization's business activities are unique.

11. *Perform E-Business in a separate company.*
    This is one of our four E-Business strategies. This may get you up faster, but it carries a high price in coordination and redundancy as well as in employee morale.

12. *Abolish the organization.*
    What if you did not consider the activity or E-Business? Instead, you just wiped out the organization. Could the work be entirely automated? Who would carry out any leftover parts that automation could not address? How would it be done? This type of radical trigger helps to define the boundaries of business activities.

13. *Split up the activity.*
    Can the activity be divided among surrounding activities? This tests your knowledge of the surrounding activities. It also tells you how strong the organization and infrastructure are for different activities. When you consider splitting up an activity, you identify the core of the activity that cannot be compromised.

14. *Cut out the paper and forms.*
    This trigger assumes that all paper and forms in the activities are questioned. Eliminating paper and forms supports the analysis of simplification, automation, and work elimination. In E-Business you want to do this, but you may not be able to for traditional business.

15. *Automate all of the business activity.*
    While certainly a dream with E-Business, the reality is that there will still be some manual exceptions. Determine how much of the activity can be automated. The parts that would remain manual are typically what is important. Automating the entire activity is not feasible for most activities. Some areas that cannot be automated include those in which customer or supplier contact is important, complex and changing logic is involved, or management decisions are required.

16. *Perform the work with different people.*
    The people who are carrying out the work may be the problem. If you could replace them, how long would it take to train new people? Are documentation and procedures sufficient to support the new staff? What part do personality and personal relationships play in the performance or lack of performance of an activity? This trigger is important because many activities depend on key staff members who have worked in the organization for many years. A small core of people may be all that keeps the existing work going. In E-Business you might use different people to support a help desk and customer service for customers.

17. *Move the work to suppliers and customers.*
    This is a goal of E-Business. However, not all activities are suited to this trigger. Activities such as order entry and accounts payable are appropriate. When you consider moving work out of the organization, define the maximum that can be moved without loss of control. Even if you do not move it, you have at least defined the simpler tasks and those that are generic. A successful example of this has been in retailing; firms like Wal-Mart have taken this very approach.

18. *Change the policies governing the business activity.*
    This trigger is tried by only a few. Policies are often taken for granted, yet they can be altered or dropped with the stroke of a pen. You want to be able to see the effects of policy change and elimination. This allows you to see the cost of the policy in terms of the activity. It also permits you to see how work may be simplified if the activity is dropped. In one project, a savings of 30% of the staff was demonstrated. Federal government agencies have begun to do this in terms of eliminating and simplifying policies, but much remains to be done. In E-Business you will probably have to modify policies to eliminate exceptions.

19. *Break down the business activity.*
    Can the activity be simplified by creating many smaller steps? By breaking down the business activity, you may be able to reorganize the work. This trigger can validate the simplicity of the activity. If you can simplify the work, you may require a less specialized staff for handling mundane parts of transactions. For E-Business you would simplify the system development. In the case of a bank that implemented E-Business, the loan application process was simplified. The detailed workflow was broken down into finite chunks and reassigned. Lower-level staff can now perform more of the work. Some of the application processing became automated.

20. *Open up information to all.*
    In some activities, people keep information to themselves, particularly when several projects or products compete for the same resources. By opening up the information, you depoliticize the approach. You can see the impact of politics and hidden information about the business activity.

21. *Have fewer people or one person do it all for accountability.*
    This is the trigger of the single point of control. If you consider having one person do everything, you see where the activity is complex and where specialized knowledge is required.

22. *Change the location where the work is performed.*
    This is an infrastructure trigger. The location where the work is performed is important to many activities. In the Crawford example, centralizing application processing increased control and reduced personnel problems. It also flattened the organization. In other cases, moving trans-

actions closer to the customers was considered. For E-Business you may
want to centralize operations for economies of scale since substantial
volume will be moved to the web.

23. *Minimize or maximize customer contact.*
    This is a goal of E-Business. The customers perform as much of the
    transactions and work as possible. An ATM machine is a good example
    of this. On the other hand, when you maximize customer contact, you
    emphasize customer service. A service-oriented department store is an
    example. In E-Business you seek the maximum customer contact through
    the web. However, you seek the least human contact that drives produc-
    tivity down.

These actions can be tested against the data and analysis that you developed in
the previous chapter. You can build a scenario out of pieces of the results of the
impact of actions.

You can proceed to either define the E-Business transactions and workflow and
then work on revising the current work, or you can do the reverse. The E-Business
start will give you more revolutionary change for the current business activities.
Starting with the current activities will tend to result in evolutionary change.

Other Approaches for Scenarios

Other ways to construct scenarios are as follows:

- Bottom-up. Proceed from the bottom upward. That is, take the information
  and detail from the previous chapter and try to improve each step and the
  transactions in the activity. This will improve steps but will leave the activity structure intact. This is a good approach for E-Business.
- Top-down. Examine the activity and create new subactivities. Logically
  move top-down in the activity to the lowest level of detail. This gives you
  greater flexibility with the activity but retains its boundaries. This can run
  into trouble in implementing E-Business since you may miss some of the
  important details.
- Infrastructure architecture. This approach builds the new business activity
  on top of the architecture that has been defined in Action 4. This approach
  will reveal gaps in the architecture and the limits of the infrastructure. It is
  tempting to define new E-Business activities based on standard e-commerce
  software. Do not do it. It does not cover the scope of the business activities.
- Outside-in. Consider the customer or supplier as you move inward to the
  activity. This scenario leaves you with a clear idea of how the new activity
  would work in interfacing. It gives less attention to the strictly internal parts
  of the activity. This is obviously a good one to consider for E-Business.

- Organization-based. Using your analysis from Actions 3 and 5, begin by considering where functions in the business activities should be placed organizationally.

These methods have different benefits and will yield different results. The triggers and the above methods can be combined when generating scenarios. Remember that these are intended to get you thinking creatively about an activity group because it is sometimes difficult to think of changes by yourself.

When you have created the scenarios, look at whether each addresses an issue. If you are not sure, reconsider the scenario. Also look at how a scenario addresses an issue. Be able to verbalize how the change will result in benefits.

Validate a scenario by changing the scope and constraints of the E-Business implementation project. If the scenario is fragile and is changed frequently, or is easily replaced, it is probably not stable. Consider replacing it.

To ensure that your scenario is understandable, present it verbally to staff members who are knowledgeable of the current activity but who are not technically literate. Take suggestions about how to reword the scenario for clarity. Keep activity descriptions verbal as long as you can to get the most feedback. Once you write things down, they tend to be more difficult to change.

STEP 3: PUT TOGETHER SCENARIOS FOR AN ACTIVITY GROUP

After generating the scenarios for each business activity (E-Business and regular), distill out the support structure for the activities. For infrastructure and technology, define a list that covers all activities in the activity group. Do the same for the organization, interfaces, and other items.

Requirements for support of the defined new activities are now present. What is missing are the changes required. Compare the existing activity and the new one (as defined by the scenario). This comparison leads to a set of required actions and changes. Use a side-by-side list for the existing and new activities. Following is the bank's example for loan applications:

Part of activity	Current activity	New E-Business	Modified current activity
System	Batch with manual steps	Electronic applications and scoring	Scan applications and handle electronically
Work control	Manual by supervisor	Automated	Automated
Interfaces	Limited to old system	On-line and web	On-line
Organization	Four tiers of staffing	Two tiers	Two tiers
Infrastructure	Multiple locations	One location	One location

The comparison identifies appropriate actions that can be used as part of the implementation plan (see Action 11). The comparison is also valuable during management review. If a manager does not want to spend money for a specific change, you can immediately map it into the area in the comparison table so the impact on the new activity can be seen.

STEP 4: TEST AND EVALUATE SCENARIOS

The following approach is both inductive and deductive. It can be repeated if you are dissatisfied with the new activity. The steps can be performed quickly with the aid of the comparison tables and analysis. Only five to seven alternatives should be considered.

Step 4.1: Assess the Range of Alternatives

You have a new business activity in mind. However, the details and organization issues are undefined. Focus on the same general activity, but identify several alternative methods of support and organization. A wide range of alternatives increases your credibility with management.

Presenting several alternatives may be a good way to deal with political issues in E-Business implementation. People may not want to make more involved changes unless and until they see the benefits. If so, start with one of the less drastic alternatives.

The range of technology alternatives may be wide while the organization alternatives are narrow. In this case, focus on the organizational alternatives and keep the technology alternatives fuzzy.

Step 4.2: Eliminate Alternatives by Comparing Them with the Current Business Activity

Using the list of alternatives, you can determine whether any alternative:

- Is marginally better than, the same as, or worse than the current activity
- Cannot be implemented to meet the E-Business deadline
- Requires a technology or infrastructure investment that is too great when compared to the benefits
- Is possible only with major organization change that has been ruled out

Do not eliminate rejected alternatives. Instead, keep them in mind to show management what could happen with fewer constraints.

You can use comparison tables similar to those of the previous chapters. Samples of tables are as follows:

- New business activity steps versus existing activity steps. Table entries indicate elimination, consolidation, change, or no action. This table indicates the degree of change.
- New activity group versus existing activity group. Moving up to the group level, each row and column heading is a business activity. The entry in the table indicates the degree of change.
- New architecture versus current architecture. This is the extent of change in the technology and systems between business activities. This gives you an idea of the effort and time required for the implementation.
- New infrastructure versus current infrastructure. This indicates the effort and time required for different alternatives.

Step 4.3: Associate Alternatives into Sets and Compare the Alternatives

Choose a critical activity for E-Business. To evaluate alternatives, you can use comparison tables similar to those shown earlier comparing the new activity to the current one. Define a set of alternatives based on a common infrastructure, technology base, or organization approach. Another set of alternatives could be based on on-line systems.

Here are some guidelines for grouping alternatives into sets:

- An alternative can belong to more than one set.
- Associating alternatives for different activities into the same set will often help in the evaluation.
- The sets of alternatives provide clarity and focus to the overall E-Business implementation effort.

Step 4.4: Eliminate All but One Set of Alternatives

To choose which sets to eliminate, look for those that involve great expense, infeasible organization change, unproved technology, or excessive time requirements for implementation of change.

Step 4.5: Select the Desired Alternative

Alternative activities in the remaining set will have common characteristics. If you have difficulty selecting one, you may want to generalize and combine them. This means a less-detailed alternative but greater flexibility in implementation. To perform the final evaluation, use two detailed methods: comparison tables and activity simulation.

STEP 5: BUILD COMPARISON TABLES

Comparison tables useful in final selection of an alternative are as follows:

- New activity group versus new architecture. This is the fit of business activities to the elements of the architecture. A lack of fit indicates that limited benefits are offered by the new architecture.
- E-Business versus regular business activities. The rows are the new E-Business activities and the columns are all activities. The entry is the relationship between the specific E-Business activity and the regular activity. This is a major opportunity to show linkage between E-Business and regular business.
- New activity group versus new organization. This table indicates the degree to which the organization changes support the new business activities. Lack of fit indicates that the solutions to different activities may not be compatible.
- New activity group versus activity issues. This table indicates the degree to which the issues identified earlier are addressed.
- New activity group versus activity characteristics. This table indicates the degree of similarity between solutions.
- New activity group versus new infrastructure. This table indicates the degree to which the new business activities depend on the same elements of new infrastructure.
- New activity group versus business objectives. This table indicates the degree to which the new activities support the achievement of business objectives.

You can use activity steps as opposed to activity groups to get a more detailed and refined picture. Again, if several alternatives appear to have the same scores in the tables, combine them.

STEP 6: SIMULATE THE NEW BUSINESS ACTIVITY

Use software to input parameters that describe the business activity. The parameters of many simulation models are as follows:

- Each step in terms of whether it is an activity or a decision
- Duration of each step and its statistical distribution
- Relationship between activities in terms of dependencies
- Arrival rate of units or customers into the system
- Cost parameters associated with doing work

Figure 9.2 shows what should be considered when identifying cost parameters. The output of the simulation model is typically the output volume, the rate of

- **Infrastructure**
  - Capital
  - Implementation of change
  - Office layout/furnishings
  - Facility moves—openings/closings
- **Staffing**
  - Hiring/termination
  - Training
- **Business activity**
  - Development
  - Training materials
  - Procedures
  - Policy development
  - Training in procedures
  - File conversion
  - Training in policies
  - Audit and control
- **Technology**
  - Hardware acquisition/installation
  - Network acquisition/installation
  - Software package acquisition and installation
- **Software tools and system software acquisition/installation**
  - Interfaces between old and new systems
  - System development
  - Data conversion
  - Integration
- **Organization**
  - Disruption caused by the move to E-Business

Figure 9.2 Sample Cost Elements

output, the costs associated with the model, the busy and idle time at each step, and resource consumption and utilization.

Infrastructure and organization change must be reflected in the parameters of the simulation. All simulation models make statistical assumptions. Verify that your situation meets these assumptions. When you consider simulation, first construct and run a simulation of the current business activity. After validating this, consider the new business activity for both the standard and E-Business cases.

STEP 7: DOCUMENT THE SELECTION

Create a flowchart of steps in both new and existing business activities. In the second step, show how the other activities in the group relate to the critical business activity. Comparison tables that reveal similarities are useful here. You also may wish to demonstrate how improving two of the activities results in synergistic benefits.

Prepare a list of the changes and improvements that are required in the infrastructure and organization to support E-Business implementation. To show why these are necessary, use some of the comparison tables that relate the activity group to the infrastructure, automation, and organization. For backup and detail, prepare a table of steps (for the critical business activity) versus each of the changes and improvements.

STEP 8: MATCH TYPES OF CHANGE TO A TIME HORIZON

The time factor will be considered during strategy and implementation. However, at this point you may want to develop a table of the changes to be made (rows) and the feasible time horizon (columns). Short-term changes will involve procedures and workflow alternations, intermediate changes will concern organization adjustment, and long-term changes will concern substantial systems changes. The table entry is the degree of feasibility for the specific time horizon. This helps provide realism and assists in defining resource requirements over time. Indicate how E-Business will be phased in through this table.

STEP 9: MAP THE RESOURCES REQUIRED TO IMPLEMENT CHANGES IN THE VARIOUS STAGES OF THE BUSINESS ACTIVITIES

Identify exact changes and list the resources that will be needed for each stage. This level of detail may show areas in which you need to rethink the new business activity.

STEP 10: DEVELOP A FUNCTIONAL ORGANIZATION TO SUPPORT E-BUSINESS

This step is often applicable to marketing in E-Business implementation. In defining a new business activity, also define the functions to support it. Don't feel restricted by the current organization. Identify the roles and responsibilities that would best support the new activity.

The management presentation for the new activity is crucial. It is discussed in depth later in the marketing chapter. Your presentation should follow this outline:

- Review the problems with current work and need for change
- List the benefits of the new business activity
- Present the new business activity that won in the evaluation
- Present other alternatives considered and a summary of analysis
- Present recommended next actions

Note that you reinforce the need for the new business activity first. Then, rather than bury people in details, describe the benefits of the new business activity. These should be readily apparent from the step comparison. At this point, identify costs and benefits by type, but do not provide an exact amount. Wait for management feedback before you flesh out the details.

Here are examples of benefits in the major areas you will want to cover:

- Customer/supplier view
  — What customers and suppliers can do on the web
  — How sample E-Business transactions can work
  — Range of information available
  — Benefits to the customer or supplier
- Position with respect to the competition
  — Products and services offered
  — Range of business activities supported in E-Business
- Infrastructure
  — More friendly and productive offices
  — More efficient offices
  — More reliable telephone equipment
  — Safer and more secure workplace
  — Lower maintenance
- Staffing
  — Different structure for E-Business
  — Reduced numbers
  — Reduced levels
- Business activity
  — Greater automation for E-Business
  — Less redoing of work
  — Lower error rate
  — Improved sales
  — Fewer returns
- Technology
  — Increased integration
  — High-cost technology replaced by lower-cost technology
  — Improved reliability
  — Lower maintenance
  — Greater flexibility to make changes

- Organization
  — Fewer management levels
  — Reduced support staff

Next, show how the new business activity will alleviate the problems with the current business activity. Then describe in detail how this will work and present other alternatives. End the presentation by defining the next steps for action.

How you document the above work depends on the style of the organization and its practices. Base any documentation and presentation on the flowcharts and comparison tables. Marketing and presentation of a scenario to management and staff are addressed in Action 10.

E-BUSINESS EXAMPLES

RICKER CATALOGS

Ricker Catalogs designed their new E-Business activities for sales at the same time they did internal activity change and improvement. This ensured that the activities were consistent. Ricker generated about four different alternative activities for sales.

MARATHON MANUFACTURING

Marathon spent a great deal of time and effort in defining their new activities. They could use their recently streamlined workflow for sales and customer service. However, the wizard for bidding and lessons learned were new activities.

ABACUS ENERGY

Abacus developed several alternatives in terms of the degree of going into E-Business for groups of purchasing and contracting transactions. In the end these all proved useful since they allowed Abacus to evolve into more and more E-Business activities.

CRAWFORD BANK

Crawford Bank considered almost an entirely E-Business loan application and servicing alternative at the start. As they went into the analysis in this action, they found that it was just too ambitious and would take too many resources and time to roll out.

E-BUSINESS LESSONS LEARNED

- **Consider the ripple effect from a change to a business activity.**
  Assuming you make the proposed changes, what effect will these have on subsidiary activities? Once you have a list of these, reconsider or rework the changes that will cause problems.
- **Make sure that people's habits and behavior fit in with the new business activity.**
  Do the staff members have the flexibility to learn new technology? Before you go to management with the scenario, test it on the employees and get reactions. Ask for specific responses to start. Ask, for example, how the new activity compares to the current one.
- **Always consider possible objections to any change.**
  Ask yourself why no one thought of the changes before. What is different now? What is unique about the new activity? Answering these questions can help you to counter arguments against your changes.
- **Take into account what the new business activity will do to the existing power alignment between affected departments.**
  With a scenario defined and organization support determined, power typically changes. Some groups win and some lose. For example, as automation increases, the information systems group increases its power. Power shifts should be part of your review of the scenarios.
- **Be prepared to deal with rumors.**
  Rumors will spread as to alternatives being considered. You can counteract negative effects of rumors by involving people in the work of change. Ask people to review the scenarios. State that you welcome feedback and ideas.
- **Identify the natural allies and enemies of the new activity scenario.**
  To determine this, think about the scenario from the point of view of the managers. Then consider what changes in organization and infrastructure are threatening to them. Work on positioning the scenario to appeal to their self-interest.

WHAT TO DO NEXT

1. Decide who will do the work. For each activity step, construct a table showing this information: P, performs the step; O, owns or is responsible for the step; and E, involved in the step on an exception basis. (Leave the space blank for no involvement.)

2. With the steps in the new business activity defined, construct a table of exceptions and normal work (rows) versus steps (columns). Place an "X" if that step is performed for the particular exception. How an exception differs from normal work and from other exceptions can be seen from the table. Compare this table to the table in Step 5.

3. Decide which new technology and parts of the new architecture (from Step 4) apply to each activity step. The rows are the elements of the new architecture and the columns are the steps for the new business activity. In the cells within the tables, enter the following codes: C, critical; B, beneficial but not critical; and N, not used. Now compare this table to the table in Step 5.

4. Determine which issues are addressed by the steps of the new activity. Put the issues in the table as rows with general business issues at the top. Then for columns enter the steps of the business activity first. Add more columns for policy, organization, and infrastructure. The entry in the table will be from 1 to 5 (1 means that the issue does not apply to the step; 5 means that the issue is addressed by the step).

5. Compare two candidates for the new business activity. How are the business activities different? The differences may involve policies, technology, infrastructure, or organization. In the rows, enter the policies, infrastructure, and organization involved. In the columns, enter the activity candidates. The table entry is a comment that highlights the characteristics of the row and column. Make a second table that is a comparison of steps. The rows are the steps and the columns are the activity candidates. The table entry is a comment about the step for that candidate.

6. For several alternatives, construct a radar or spider chart that compares alternatives in terms of the following: risk, duration of implementation, tangible benefits, overall benefits, estimated cost of implementation, organization change required, technology required, and size of implementation effort.

7. List the business activity generator ideas as rows. In the second column, give the results of using these ideas. Comments and notes appear in the third column.

Part IV

Prepare for Your E-Business Implementation

Chapter 10

Action 8: Define and Measure E-Business Success

INTRODUCTION

E-Business implementation involves benefits and risks. Some of the major benefits have been discussed. Here are some of the risks:

- The E-Business site does not attract new business. Instead, current customers use the web to get lower prices. The web has cannibalized the customer base. Alternatively, suppliers continue to use fax machines and the telephone, killing off potential productivity gains.
- The E-Business site is so successful that some combination of the business activities, technology, and systems fails to be scalable in terms of the volume.
- The E-Business is sufficiently successful that the employees feel threatened.
- Your strategy toward E-Business was based on some use of the current activities. This did not work and customers are screaming about bad service.

These problems have a habit of becoming widely known. E-Business customers and suppliers tend to be a vocal group. They will talk about these problems in chat rooms. In some cases, if they get sufficiently angry, they may set up a web site that attacks the firm.

It is not surprising that managers want to know the extent of risk. People seek benefits. Did they get them? Measurement of the old and new activities gives an understanding of what is happening. Measurement in E-Business is more complex because the measurements are wider. You also must compare the current and new version of the traditional activity as well as the fit between the E-Business activities and the new standard business activities.

A second reason measurement is important is that it allows you to control the new E-Business activities. By measuring an activity while it is in operation, you can affect its direction and behavior. You also measure to support the decision-making approach regarding E-Business on an ongoing basis. More to the core of E-Business, measurement is critical to keeping the web site competitive.

In standard business activities, the problem is usually collecting the data. While this is true in some areas of measurement of E-Business, there is also the opposite problem—being overwhelmed by data. In the bank example for loans the bank purchased software that tracked each mouse click and operation performed by a customer. How would such enormous amounts of information be employed? In general, its only use was to refine and simplify web site navigation.

When choosing a measurement method, address the following areas:

- Data collection. Can you collect the data in an automated and consistent manner that will admit standards and comparison over time?
- Efficiency. Is the measurement method cost efficient so that the cost of gathering, analyzing, and reporting is reasonable?
- Flexibility. Is the measurement method flexible enough to allow detailed, ad hoc questions to be addressed? Can the measurements be adjusted to handle more work?
- Verification and validation. Are the measurements capable of verification? Are they sufficiently complete and of acceptable quality?
- Interplay with decisions. Can you isolate the measurement method from management decisions? If you cannot, a management decision may affect future measurement.
- Activity effect. Is the measurement method unobtrusive, so that measurement has no negative impact on the business activity?
- Perceived value. Do all of the people involved in collection and analysis receive some value from the measurement? If not, the measurement method may fall into disuse.
- Support for ongoing measurement. E-Business depends on this. Measurements have to be affordable to be able to be done repeatedly.

INFORMATION DECAY AND THE PRESSURE OF TIME

As information ages, its value and use decrease. You can see this problem in military activities in which movement is rapid and change is quick. The same pressure exists in E-Business. If errors occur in a critical activity, decisions could be incorrect or the timing could be off. Operational impact is possible with angry customers. This happens in standard business activities too. For example, if year-end accounting was incorrect and the errors were undetected, it may be impossible to reconstruct the data within a reasonable time. In E-Business, you focus on e-critical activities, so measurement and analyses should be performed in a timely manner.

THE MORE YOU COLLECT, THE MORE THEY EXPECT

Given the potential problems arising from insufficient information and analysis, you might tend to err on the side of collecting more data. This is possible in E-Business. If you collect more information, people employ more of their time to analyze the information and then provide you with summaries. People need feedback, so they will expect you to return with results that use the data they provided. If you do not use the information, you lose credibility. People will become sloppy or less willing to collect the information next time, believing that it will not be used.

MEASUREMENT'S ROLE IN DISASTER RECOVERY AND BACKUP FOR E-BUSINESS

E-Business depends on technology. There must be backup and recovery capabilities for transactions. The measurement method is important to disaster recovery, a major area of data processing. All major computer installations have disaster recovery plans for critical systems, whether the disaster is natural or man-made. Business resumption planning was developed to address the recovery of the general business activities—staff, files, procedures, and other resources. Measurements are one of the triggers for business resumption and disaster recovery.

TRADE-OFFS BETWEEN VALUE AND IMPACT/EFFORT

Measurement is an overhead activity; it does not yield the end product of the business activities. Detailed measurements can disrupt the work and lower productivity. Thus, there is a trade-off between the value of measurement on one hand and cost, time, and effort on the other.

To justify the cost and effort of measurement, consider who will be receiving the information gained from the attempts at measurement. These audiences include the department staff, the supervisors, the managers, and outsiders. You want to be proud of the performance of the E-Business activities.

Consider the lending example with the bank. Its implemented measurements were as follows:

- Use of the web site by customers for application preparation. This included the volume, system performance, and customer satisfaction with the web site and the application process.
- Staff performance in terms of volume of calls, results of calls, and activity per employee. This information helped to measure and compare employee

performance as well as overall performance. This also allowed for the measurement of activity performance on an individual level.

- Aggregate staff performance in terms of total volume of calls and work, number of busy signals on incoming calls, and application processing activity. This data used the individual production results to assess an office and group. It also gave a total production picture, including the cost of operations. A comparable measure existed on the web in terms of response time and failure due to overload.
- Quality and results. In loan applications and servicing, this included the extent of electronic use, labor savings, fixing errors, and customer service.
- Performance and results ratios. These combined activity performance data with information on the results of the banking activities.

Now consider the various audiences. What does each derive from the measurement?

- Customer. Adequate service and performance means that the customer will not only continue to use the web site, but also probably refer other people to it.
- Department employee. Feedback on performance, rewards for good performance, and indication of areas of improvement.
- Supervisor. Statistical data on volume and performance by employee and by the group over time, along with a comparative analysis.
- Management. Summary performance reports, cost, and performance comparisons with other institutions. In addition, there are relative performance measures relative to other banks.
- Outsiders. How the bank performs in E-Business impacts the value of the stock and how the company is viewed in the investment community.
- IT organization. IT is concerned about reliability, availability, flexibility, and support requirements.
- Marketing. Marketing is concerned with not only today's performance, but also the ability of the business activities and systems to support new products and promotions.

CLASSIFICATION OF MEASUREMENTS

Measurements can be classified in a variety of ways, including the following:

- Frequency. Is the measurement regular, periodic, or ad hoc?
- Level of detail. To what level of detail will E-Business be tracked? The transaction and the customer are two options.
- Black box model. Total results are measured; no examination is made of the detailed performance or the internal workings of the activity.

- White box model. Information is collected on the activity and the steps in the activity with automated tools. White box analysis and testing allow the data from inside the activity to be combined with the results. This supports the analysis of more detailed issues.

Black box methods were at one time the only possible approach. White box measurement was restricted to direct observation and study. Using the bank example, before it implemented the new activity, the bank employed a black box approach for loan application measurement. Technological support through the E-Business software, combined with management's desire for information, provided the impetus to implement a white box measurement approach in the new activity.

With so much data from the web and so many software analytical tools available, there is a tendency to implement white box methods right away. Do not do this. You will be swamped with data. You will have many other issues to address. We suggest that you use summary, black box measurements for customer and supplier activity. You can move to white box measurement later.

The purposes of measurement are to get support for the continuation of the E-Business project, to measure the progress of work on the project, and to determine whether benefits were achieved from the investment in the E-Business project. Measurement can also relieve much of the doubt and worry that are common during E-Business projects for the following reasons:

- The company has invested a great deal of money and image in E-Business.
- Some managers fear change. Many know the buzzwords of E-Business, but do not really understand it.
- People have doubts about the value of the effort put into the standard business and E-Business activities.
- People have observed past failures in E-Business.
- E-Business changes tend to take place over an extended time.
- Employees still rely on certain key activities for their livelihood.

The scope of E-Business measurement includes all parts of the project—from initial conception and selection of business activities to the postimplementation review of results. Measurement also is a regular part of E-Business and regular business.

MILESTONES

The major end products are measurements and a database of supporting measurement information for later analysis. Here are some end products for E-Business implementation.

- Measurement of the E-Business implementation project itself
- Performance of the systems and technology to support E-Business

- Measurement of the economic success of E-Business
- Impact of E-Business on the other activities, organization, and policies of the firm
- Marketing measurement in terms of customers and suppliers reached as well as flexibility of the activities and systems
- Measurement of the activities to support E-Business volumes

METHODS FOR E-BUSINESS

STEP 1: DEFINE THE MEASUREMENT STRATEGY FOR E-BUSINESS

Such a strategy encompasses the following:

- Identification, collection, analysis, presentation, and decision processes for normal situations. This applies to the current business and E-Business activities.
- Approach for measurement in the event of abnormal conditions such as high-volume periods. While an advanced prediction is not possible, you could hypothesize situations in which you would need more detail or in which a specific issue relating to budget, quality, or resources arose.

At the bank, the traditional normal measurement method consisted of a series of summary reports and on-line access to a database of summary information. This information was provided for transactions above the level of an individual loan. For exceptions, the method yielded detailed information that allowed management to "drill down" to individual loan level data. This was successful for supervisors and middle-level managers. It was inadequate for marketing, management, and IT.

Ricker Catalogs collected a large amount of information, including:

- Number of customers by profile who used the site versus the paper catalogs
- Average purchase and number of items by customer type
- System performance in terms of response time, reliability, availability, cost, and throughput (volume of transactions)
- Fulfillment performance in getting goods to customers
- Customer service contacts in terms of total volume, volume by type, complaints, etc.

Marathon Manufacturing created a value-added web site that included wizards for bidding, lessons learned on setup, maintenance, and use of equipment, and customer service. In addition to the Ricker types of measurements, Marathon tracked the use of these additional features as well as the growth in the scope of the web site.

Abacus Energy measured the internal staffing in contracting and purchasing as well as volume characteristics for suppliers using the web site. They also employed measures corresponding to Ricker's.

Measurement of the E-Business implementation project is important because you want to extract lessons learned and use these for managing future changes and growth. You also want to measure it because you have substantial risk and investment at stake. This measurement goes beyond schedules and costs. You want to know if you are going to achieve the anticipated results. Otherwise you may risk not detecting a problem until it is too late—after people in the department have labeled the project as a failure. Use the following for measurement information:

- Normal project management methods for summary information about the project
- Detailed data collection and analysis method to probe into the E-Business implementation project in depth
- Ability of the E-Business project to deal with change during the project
- Capability to handle issues that arise. In E-Business these can be either problems or opportunities

STEP 2: DEFINE THE MEASUREMENT CRITERIA

Some common measurements are as follows:

- Revenue
  — Gross revenue generated by the web
  — Profits from web sales
  — Comparative costs of web versus traditional business
- Customer and supplier satisfaction with the web site
  — Extent of use of the web
  — Trends in the use of the web
  — Demographic information
  — Extent of repeat business
- The web site itself
  — Ease of navigation around the site
  — Extent of information available
  — Number of products or services available
  — Flexibility in handling change
  — How the web site stacks up against the competition

- Cost
  - — Operational cost of production use
  - — Effort to maintain, enhance, and support the systems
  - — Capitalized cost of equipment and other components
- Performance
  - — System response time
  - — System throughput—volume of work per unit time
  - — Error rates
  - — Down time and time to repair/recover
  - — Number of simultaneous users on the system
  - — Availability of systems and technology for work
- Other
  - — Turnover of computer support staff
  - — Assessment of documentation of systems
  - — Age and obsolescence of technology and systems
  - — Number of original programming staff with application software
  - — Size of system in terms of total program size and number of programs
  - — Size of the databases and files
  - — Response time to implement an enhancement

Infrastructure and Technology

Measuring the infrastructure is situation-dependent. However, some common examples include the following:

- System performance
  - — Response time for customers or suppliers
  - — Ability to handle peak volumes
  - — Reliability
  - — Availability of the site
  - — Quality and accuracy of the information on the site
- Cost
  - — Cost to add new information and features
  - — Operation and maintenance cost
  - — Installation, upgrade costs
- Physical attributes
  - — Number of locations
  - — Staffing per location
  - — Systems and technology
  - — Stability of operations facilities
  - — Availability of facilities
  - — Safety record of facilities

Organization and Staff

Go beyond simple costs and head counts. Here are some examples:

- Cost
  — Direct staff cost
  — Management cost
  — Overhead and burden costs
- Noneconomic measurements
  — Stability of organization—frequency of organization change
  — Staff roles, responsibilities, and levels
  — Employee turnover
  — Employee sick-outs
  — Staff seniority
  — Performance evaluations of staff morale
  — Number of employee problems
  — Time and effort consumed in training

Measuring Interfaces

Interfaces between and among activities and systems are an important measurement source. Many errors and problems with systems and activities arise in the interfaces. You can change a business activity, but if you do not address the interface, you are still likely to have problems. E-Business transactions tend to cross multiple systems and departments. Therefore, measure an interface as follows:

- Response time of the interfaces
- Timeliness of interface for supporting E-Business
- Amount of work required by the interface
- Error rate of work in transactions using the interface
- Time and resources required to support the interface
- Error rate of the interface
- Availability of the interface

Following is a summary of a few of the measurements:

- Type of work
- Volume of work by type
- Overall time for processing transactions
- Error rate by type
- Redoing work and error correction time and effort

For all of the components (infrastructure, systems and technology, etc.) you will collect information over time on a recurring basis. You could collect some information at lower levels of detail down to each item purchased on the web, but this may be prohibitive to analysis in a timely manner.

The criteria discussed above lead to what data will be collected. You may have a choice as to how the data will be collected. Pick the approach that is the most consistent and can be defended in a presentation as being objective.

STEP 3: DETERMINE THE ANALYSIS APPROACH

From the above information you can construct ratios between, and composites among, categories. Examples are as follows:

- Customer or supplier complaints and satisfaction surveys
- Volumes of purchases—both totals and by type
- Number and type of marketing promotions and new products supported
- Extent of new information and products put on the web site
- Cost per transaction
- Throughput per employee
- Response time per transaction
- Cost per employee of infrastructure and systems
- Errors and cost of redoing work per transaction
- Distribution of time between the system, interface, and manual parts of the activity

You can develop trends on the basis of individual cost or performance items as well as for ratios and composites.

STEP 4: DO THE MEASUREMENT

Carry out the measurement as quickly as possible. Organize the information as it is collected to ensure that there is completeness and that it makes common sense. Establish the ongoing E-Business measurement methods. Also, consider building the presentation as you go.

Measure the Current Business Activities

Why measure the current activities? Because you need a benchmark. Standard measurements are probably available. Since you are improving or replacing the business activities, it is likely that these measurements are poor, or that the current measurements are not indicative of the state of the activity, or both. Measurement of the current activity is limited by the following:

- Limited funds to invest in new measurements for the activity on a regular basis because it is going to change with the new activity
- Staff who are aware of the project and may resist or slant any additional measurements you take
- Resources are being drained by E-Business implementation, limiting the resources available to measure the current activities

You can start with the current measurements and consider the complaints people have about the activities, such as the activity being expensive or prone to errors. More complex comments might be that activities are inflexible and incomplete. For most current activities, measure the total activity. This typically means measuring the part of the activity that deals with exceptions. Often, the exception effort is separate from the measurement of the activities and skews the results. Suspect this to be so when you notice that standard measurements do not match up to complaints.

Measure the Concept for the New Business Activity

If the organization does not have an activity in place—just a concept and a design—follow up on the comments in Actions 6 and 7 to build a model to simulate or estimate how the activity might behave. The key here is not the exact data. The measurement here may indicate that a high probability exists that the new activity will produce benefits that justify implementation. Limits exist. Any information and results you derive must be comparable to the measurements of the current activity. Demonstrate that the new activity will be an improvement.

Also estimate the results of the method for different alternatives. These alternatives may involve centralized or distributed organizations and various types of technology, among others.

Measure the New E-Business Activity

In Action 7 your E-Business activity was developed to incorporate measurement. To implement it, be sure that the measurement method is in place. Begin by collecting measurements. Look for information that helps you compare results of the planned activity versus those of the old one. You also want to ensure that the measurements reflect reality. For example, in the Vietnam conflict, measuring progress through number of acres held or body counts did not work. Interviews and the number of defections, as well as the nature and mix of goods on the Ho Chi Minh Trail, were more enlightening. In narcotics interdiction do you measure success by the volume of drugs seized, the street price of the drugs, the number of arrests, or the number of people in treatment programs? These examples point to the need to validate the completeness of the measurements.

Measure the E-Business Implementation Project

Apply the normal measurements of project management, in terms of schedule and cost. To do this, determine the general perception of the project. Is progress noted? Much of the work and milestones have to do with infrastructure and design of the new business activity. Is the design sound? Is the project retreating into making small activity changes? Is the new activity beginning to look more like the old one?

STEP 5: PRESENT THE MEASUREMENT RESULTS

As you prepare the presentation, try it out on the people involved in the work and the project. The feedback may indicate if you have missed some key message from the data. For E-Business you will want to give a dry run with simulated numbers to validate that the level of detail in the measurements as well as the scope of the measurements are valid.

EMBEDDING MEASUREMENT INTO E-BUSINESS

The best way to embed measurements into an E-Business activity is to collect data automatically by the systems. This is a benefit of e-commerce software and on-line systems. Status tracking of a transaction or piece of work occurs as it flows from step to step. You can record the time, person, and any other pertinent information for each step. Use manual collection only as a fallback because it is subject to interpretation and variance.

AD HOC, REACTIVE MEASUREMENT

If an issue arises and the current measurements are inadequate, first state the issue clearly. After collecting additional data, write possible outcomes and actions. Then define further data you want to collect.

The next issue is the data collection method. You do not want to disrupt the activity. Some ways to collect additional detail are the following:

- Collect data through the system. Collecting too much data in a network can slow down the system for customers and can also result in volumes of data that you cannot cope with.
- Conduct an overall review of the activity. This may take more time, but you can collect the additional data as part of the review, and this will be less disruptive of the work. Also, issues are really symptoms. A review will reveal the causes.

- Conduct direct, informal observation of the business activity, or look at E-Business statistics on a regular basis.

MULTIPLE BUSINESS ACTIVITIES

With an activity group, there is a need for a consistent measurement approach across the group. Differences can exist in the level of detail based on the importance of the activity, but measuring even smaller activities will allow you to do benchmarking.

In the case of lending for the bank, there were eventually five lending products which employed a common set of measurements. These included personal unsecured loans, automobile loans, mobile home loans, credit cards, and home equity loans. Both application processing and servicing were supported.

THE COMPETITIVE EDGE

You will achieve success not only by the new E-Business activities, but also by the stories revealed by the measurements of them. Even with the priorities of E-Business, you are constantly competing with other projects for management support. People may be pulled off the project to work on "emergencies." This is one reason for being proactive in measurement. Another reason is that, if you constantly measure your project and the activities, people will trust what you do. There will be fewer ad hoc, disruptive measurement efforts.

Train yourself to look behind the data. Keep the dilemma of the criminal justice system in mind. Is a rise in crime due to more crime, or to improved reporting? Jumping to simple interpretations leads to the questioning of results. Next question the data. Then hold the activity itself up to question.

Search for simple minimum standards of measurement. Measurement is overhead. Automation reduces the burden, but you still have to analyze and then report on the information. It is easier to expand measurement rather than cut it back. Also try to use the measurement information to generate ideas on how to improve the activity. As long as you have the data, why not do more with it?

WHY DOES MEASUREMENT FAIL?

Areas of failure include the following:

- Collection of the wrong information. A survey team collects information on a business activity, including basic volume and work statistics. It fails to collect information on redone work and errors.
- Collection of information that is too general to be of use. You collect infor-

mation on trends on a region but do not pick up information on a specific
company in the region.
- Information gathered at the wrong time. You are thinking of investing in
  a small business. You observe the business on the weekends, notice that it
  is busy, and assume that it is a real moneymaker. During the week, however, business is slow. Be sure to gather information at a variety of times.
- Collection of too much information. Analysis of the volume becomes
  impossible.

MEASURE YOUR WORK

You want to be realistic and nonbiased when measuring. Here are some questions to ask:

- Do the measurements and observations of the activity match up with what
  people think? If not, either your measurements are in error, or you have
  a perception or political problem.
- Can before and after comparisons be made easily in terms of the business
  activities?
- Have people's expectations and goals changed? Are the measurements
  keeping up with these?

A major test is whether you are being realistic or biased in terms of measurement.

E-BUSINESS EXAMPLES

RICKER CATALOGS

Ricker knew that measurements were important. However, they did not at the
beginning assign specific staff to do the work. Management assumed that departments would be performing the measurements on their own. They were not. This
problem surfaced in a management meeting when the CEO asked about characteristics of customers on the web versus traditional catalogs. Faces were blank in
response. It was time for formal measurement. Two people were assigned to measurement full time to get things sorted out.

MARATHON MANUFACTURING

Marathon wanted measurements from the start. An entire set of reports, charts,
and graphs were prepared to simulate how measurement results would be pre-

sented. The effort in collecting and analyzing data was covered in a number of meetings. Marathon even went so far as to simulate management decision making.

ABACUS ENERGY

Abacus had a much smaller E-Business effort than the other examples. They did not place a high priority on measurement until some suppliers complained that E-Business seemed to be slower than regular business. This triggered measurements to be implemented.

CRAWFORD BANK

Crawford implemented the same measurements as for the traditional lending business. This provided useful summary information for E-Business. However, they did not consider the demographics of customers on the web. They assumed they were the same as regular customers, but they were not. They uncovered this when they wanted to expand the products on the web. It is then that they found that little data had been retained.

E-BUSINESS LESSONS LEARNED

- **Use self-supporting measurements.**
  In order for measurements to be useful over an extended period, the data collection process should be self-policing. Ensure accuracy by seeing that it is in people's self-interest to collect the information. Have a backup for each individual who is collecting, analyzing, or summarizing information. You have an edge here in E-Business since much information can be captured automatically.
- **Match measurement reports and graphs to style.**
  Give management an opportunity to have a say in which graphs and charts they would like. Show them several alternatives.
- **The measurement technique should be flexible.**
  Even an E-Business activity and its systems will age over time and grow in complexity and eccentricities. Be sure that the measurements you have selected adequately address these changes. Measurements cannot do this if it is entirely manual. Also, the measurement method must address exceptions as well as the normal activity.
- **Use measurement data to identify activity bottlenecks.**
  Measurement data are used not only to improve the activity and but also to find the next bottleneck in the activity.

- **Use accepted measurement practices.**
  See how other activities are measured. Are any of their measurements adaptable to your work? If they are, use them.
- **Continue to improve your measurement techniques.**
  Keep the data collection and gathering method constant, but continue to improve the analysis and reporting of measurements.

WHAT TO DO NEXT

1. Rate how easy measurements will be for the activity group you have chosen. Use a scale of 1 to 5 (1 is difficult or not measurable; 5 is easily measured).

Measurement	Availability	Accuracy	Completeness	Quality

2. For the systems and technology corresponding to the activities you have chosen, make a list of the measurements that you could gather and rate these according to the criteria defined in this chapter. Use a scale of 1 to 5 (1 means very difficult to collect; 5 means that it is easy to obtain).

Measurement

Systems			

3. Repeat Action Item 2 above, but rate the infrastructure. Again, start with the list developed in the chapter.

Measurement

Infrastructure			

4. Repeat Action Item 2, but rate the organization. Begin with the list in the chapter.

Measurement

Organization			

5. Repeat Action Item 2, but rate the interfaces between activities.

Measurement

Interfaces			

6. Recurring data collection is sometimes necessary but difficult. Rate each activity on ease of recurring data collection using a scale of 1 through 5 (1 is impossible to collect; 5 is easy to collect).

Criteria

Activity	Automated Capture	Analysis Effort	Manual Integration of Data

Chapter 11

Action 9: Develop Your E-Business Implementation Strategy

INTRODUCTION

With new E-Business activities defined and supported by delineated changes in the firm's systems, organization, infrastructure, and policies, E-Business implementation appears to be a daunting effort—complex both technically and politically. In traditional projects an implementation plan was built first and work proceeded from there. Many E-Business projects have floundered because of this, indicating that an overall strategy for implementation must come first.

THE NEED FOR AN IMPLEMENTATION STRATEGY

The E-Business implementation strategy identifies the scope of change in terms of organization, business activities, policies, technology, and infrastructure. It also indicates how changes in each of these areas combine into projects. The strategy identifies how you will realize the benefits and vision of the new activities.

Here are some additional reasons for implementation strategy:

- Scope. The broad scope of different projects cries out for an overall umbrella for E-Business. A vision can be difficult to relate to projects because the vision is general while the project is detailed.
- Consistency. You must ensure that projects and efforts are consistent across the organization. This is the same as having groupings of projects under the strategy within the common E-Business vision.

- Focus and sequencing. Focus is vital, especially with projects that can span a decade. The project can outlast some of the staff (as was the case with the loan processing example). People must be aware of which projects are underway and of the sequencing of the projects. Focus will help prevent management from tinkering with the E-Business effort.
- Flexibility. It is difficult to accommodate different technologies and lessons learned in a single project plan. By definition, a project plan addresses end results. Flexibility outside of the project plans can be achieved through the strategy.

In the example of the loan application processing, the strategy was to employ new on-line systems. These supported applications for various bank products. Successive products paved the way for further consolidation. This wave was followed by transitioning into servicing for new customers. As the E-Business market evolved, the strategy frequently involved retrofitting areas previously touched by improvement work.

Thus, the strategy is a relatively brief statement of the direction of implementation. Read between the lines and a number of additional factors are revealed. In the loan application example the new on-line systems would require the following:

- Several waves of organizational change and consolidation
- At least two and possibly three waves of technology modernization
- Enforced stability between waves of change
- Integrated business activity, technology, and infrastructure change

The strategy should also include the following:

- Phased approach in implementation. What are the general phases of work? For the bank this included all parts of application processing from initial editing and review to decision making.
- End products associated with each phase. Where do you conclude each phase? Process and organization change as well as systems changes were indicated in this stage for the bank.
- Measurement approach to be employed. How will you know if the strategy is successful? The measurement criteria for the bank were the tangible goals of productivity and increased volume of high-quality loans.

WHY DO ORGANIZATIONS FAIL TO DEVELOP AN E-BUSINESS STRATEGY FOR IMPLEMENTATION?

Some companies think that if they have a general E-Business strategy, then they do not need an implementation strategy. It seems obvious that an implementation strategy is useful, yet many organizations fail to develop one. Experience points to the following reasons for this:

- Fear of revealing the total scope. It is feared that if the true scope of the change is revealed, middle management and staff will become hesitant and obstruct the E-Business implementation.
- Generation of enemies. The scope and strategy reveal the direction and depth of change across the organization and systems. Managers may collude and try to resist.
- Impractical and unrealistic project. Painting a picture of an overall strategy may give people the impression that the project cannot be completed.

To address these concerns, focus publicly on a general vision of the future. Dates, timetables, benefits, approaches, and other such details in the E-Business implementation strategy should keep a low profile.

In the loan application example, it was clear that the overall implementation plan would be very broad and extend over a long period. Because the implementation included more than 6000 bank employees, crossed over 15 major systems, and affected more than one million customers, it would probably impact revenue and costs and involve considerable risk. Moreover, e-commerce software was in its early stages.

Once the bank realized the consequences of change and the fact that benefits would be cumulative as implementation progressed, development of an implementation vision and strategy became more important. Factors in the vision were improved loan volume, productivity gains, improved service, economies of scale in handling large volumes of data, reduced staff turnover, and a reduction in number of management layers.

The E-Business implementation strategy centered on changing one critical banking area. Then there would be a pause to consolidate and measure the benefits. The pace of implementation quickened as similar functions were changed and E-Business spread to other areas of the bank.

THE E-BUSINESS IMPLEMENTATION STRATEGY HELPS AVOID FAILURE

Without a strategy, the E-Business implementation may succeed initially but ultimately fail. Below are seven common reasons for failure in implementing change. Many of these reasons reflect the need for a strategy:

- A sense of urgency for change is lacking.
- A vision is lacking. The guiding coalition of management is lacking. The

implementation strategy serves to reinforce the vision and make it feasible and tangible.

- The E-Business vision is not communicated effectively. Many visions are vague. Most people prefer more tangible, action-oriented statements. Properly constructed, the strategy can assist in communicating the vision in a more palatable form.
- Obstacles to the E-Business vision are not removed. The implementation strategy can serve to point the way in identifying obstacles early.
- The systematic planning and creation of short-term wins is lacking. There are short-term or quick hit gains by changing current activities to eliminate exceptions and workarounds. The E-Business implementation strategy helps to provide the sequencing of projects and indicates order of delivery.
- Victory is declared too soon. The E-Business implementation strategy serves to keep long-term goals in mind. It can head off declaring victory prematurely.
- Change in the corporate culture is not achieved. Sometimes, it is difficult to see how one project or a set of unrelated projects can change the culture. The E-Business implementation strategy can serve as the bridge between the E-Business vision and the organization culture.

Additional problems that follow from the lack of a strategy are the following:

- Ad hoc changes in the E-Business implementation effort occur.
- Vendors and individual managers may seek to fill the vacuum and take over the project. Their own strategy can then support their own tactics.
- Changes to the project may appear to be signs of disorganization. A strategy that supports change and evolution produces less surprise when change occurs.
- Without an E-Business strategy the project can become reactive. Project leaders may spend too much time reacting to change and suggestions.

EXAMPLE OF DISASTER: INSURANCE COMPANY

An appropriate E-Business approach was developed for a large insurance company. The purpose was to implement a new business activity for insuring and servicing automobiles, homes, and mobile homes on the web. The work included evaluating insurance applications, issuing policies, collecting premiums, and handling claims. The goal was to establish one consistent business activity that was valid for all types of collateral (autos, houses, etc.) with as much on the web as possible.

The company started with changing the homeowner's insurance area, but there was no strategy. As E-Business marketing opportunities surfaced, management changed direction and priorities—five times in three years. The project responded to management each time. The entire project failed and was terminated. Costs were in excess of two million dollars and nothing was implemented.

An E-Business implementation strategy provides a structure and framework for the detailed plan and implementation to follow. The strategy will help to overcome resistance to individual projects and also will enable a better understanding between staff and management. A suitable implementation strategy should permit flexibility, allowing for change in scope of the implementation plans.

The scope includes all aspects of the business activities, new and old, as well as business, infrastructure, and organization changes. Many mistakenly include only the infrastructure and activity changes within the implementation strategy. People then get the impression that when these are completed, the project is finished. At that point it is difficult to rewrite the strategy and gain acceptance of the new version.

MILESTONES

The major end product is the E-Business strategy. As with the end products in previous steps, the strategy is supported by comparison tables. Make sure that the strategy is understandable and tangible to everyone. You will be referring to the strategy often to resolve issues during implementation.

METHODS FOR E-BUSINESS

STEP 1: DEFINE THE DIMENSIONS OF THE E-BUSINESS IMPLEMENTATION STRATEGY

Technologies, systems, and business activities are dimensions that come immediately to mind. Organization, infrastructure, and measurement are also key dimensions. Your strategy should address each dimension. The omission of one or more may limit the scope and flexibility of an E-Business effort.

The E-Business implementation strategy is the road map and link between the E-Business vision and the detailed implementation plans. The dimensions of the implementation strategy must match both the scope of the vision and the type and nature of the detailed implementation project plans.

When developing your strategy, address the six dimensions in the following manner:

- Technology. The ways in which technology supports the activity change should be part of the strategy.

- Systems. Address what will happen to the current application systems and new systems.
- Business activities. Highlight policy areas to be changed in the business activities, procedures, and policies.
- Organization. Use general terms to describe change in this area
- Infrastructure. This area absorbs most of the expense and project lead time, so give this area sufficient attention in the strategy.
- Measurement. Define the criteria for measuring success.

You could view these dimensions as the rows of a table. The format is shown in Figure 11.1. Columns in the table are phases. The entry in the table is the strategy for that specific dimension and phase. Such a chart shows how the implementation strategy crosses the phases and helps in showing management the overall strategy. Consider using a four-phased approach. You might want to use the following:

- **Phase I—Prototype new system, begin work on infrastructure, and establish the marketing structure for E-Business.**
  In this phase initiate work on the system and the activity, and establish the infrastructure to support widespread deployment of the new business activity and system in a later phase.
- **Phase II—Pilot the new business activity and refine the system; complete the infrastructure.**
  In Phase II, complete the infrastructure preparation. It defines the new activity in the context of the new system (the pilot). Test the combination of the two in this phase.
- **Phase III—Complete and implement the E-Business activities and measure the results; begin organization change.**
  Following success in Phase II, spread the business activity and system

Strategy Areas	Phase I	Phase II	Phase III	Phase IV
Activities				
Systems				
Infrastructure				
Technology				
Organization				
Measurement				

Figure 11.1   Implementation Strategy Table

through the organization. With this near completion, pay attention to organization change.

- **Phase IV—Complete organization change; begin the measurement method.**
  When the E-Business implementation ends, implement measurements.

You can prepare a table where the rows are business activities, systems, infrastructure, technology, organization, and measurement. The entries in the rows and columns of the table consist of the projects appropriate for this cell. Note that relationships exist across rows as well as between columns.

STEP 2: DEFINE ALTERNATIVE E-BUSINESS IMPLEMENTATION STRATEGIES

Many alternative strategies must be considered. They should address what to do and how to do it. Following are several types of alternatives.

Revolutionary Type of Implementation Strategy

In this approach, the implementation strategy addresses change to many business activities at the same time. For example, you may want to change the activities and organization concurrently. This approach is typical in radical business process improvement literature. Revolutionary strategy has advantages, such as faster completion time and less general disruption over the entire time, but it also carries risk. One disadvantage is that the approach ignores the learning curve from an incremental approach. Also, with so much change, projects may get in each other's way. In some instances this is the preferred approach for E-Business.

Infrastructure-Based Implementation Strategy

Examples of infrastructure-based strategy are relocation, distribution of departments, and consolidation of functions. Activity changes here take a backseat to infrastructure. Many E-Business projects fail as a result of some aspect of the infrastructure.

This is a viable approach for large projects, though it can delay benefits until infrastructure changes have been completed. The most common reason for using this approach is that the infrastructure is preventing the current business activity from functioning and is inhibiting the implementation of the new E-Business activities. However, it is dangerous for E-Business to be based on infrastructure alone.

Organization-Based Implementation Strategy

This is a classic implementation strategy. The approach is to change the organization first. With new managers and staff in place, attention moves to the work. Like downsizing, this can force savings, but the total cost may be high. Morale and productivity can suffer. People with a working knowledge of the business activity may leave. For people who are new or who remain, the learning curve can delay the E-Business implementation.

Incremental Mixed Implementation Strategy

In this implementation strategy you gradually change the business activities, technology, and infrastructure. At a later time, with sufficient implementation, you can attack organization issues. This approach offers the advantages of minimal or no disruption and little risk. The drawback is that results are achieved at a slower pace. In some large organizations, this may be the only viable approach. This is probably the most suitable and comprehensive approach for E-Business.

STEP 3: DEVELOP THE E-BUSINESS IMPLEMENTATION STRATEGY

Use the tables constructed earlier to develop the E-Business implementation strategy. Following are a series of steps to develop the strategy:

- Create the rows of the table described earlier. For the infrastructure area divide the work into two or three phases. Each phase should result in usable, tangible parts of the infrastructure. Keep the time period of each phase vague.
- Move to systems and technology. Now place the major development areas into the phases you have defined. The development areas include prototyping, design, development, conversion, testing, and integration. Development should finish in the last phase of the infrastructure.
- Consider the business activities. Overlay the activity, policy, and procedure changes and their implementation on the phases. Completion of the new E-Business activities should occur shortly after that of the system.
- In the area of organization, you may wish to create a separate, later phase for organization change.
- Measure the current business activities in the first phase. The new measurement method will be established in the last phase.

To enter activities across the phases in the table, work with each row separately. Each activity will probably consist of a small project. You now have a table of sectors of E-Business versus phases. To check your work, first ascertain whether there are any dependencies that you failed to identify. To adjust dependencies, move activities between phases. Carefully review the tangible results achieved at the end of each phase and identify the benefits of each phase. Write these below each column of the table. In general, the more dense the table (the fewer empty cells), the more parallel effort is possible.

Next, identify subprojects needed for each row activity. The projects and subprojects will become the center of the implementation plan. Write the resource requirements at the bottom of the table, so that you can tabulate them.

You now have a general phasing approach for the elements of the implementation strategy. The nature of the activities in each phase identify the "what" of the strategy. Enhance the strategy by providing a statement of how work will be done. Do this for each row and column.

STEP 4: EVALUATE YOUR E-BUSINESS IMPLEMENTATION STRATEGY

Your implementation strategy will be tested and tried, and may be changed during implementation. How will it hold up? Examine how work in the rows and columns can be altered, or shifted, to respond to the pressure or crisis. Following are some examples of change and impact:

- **Reordering of priorities**
  Reordering business activity priorities changes the rows for systems and for the activity. There is typically less impact on the other rows. Thus, you should be able to create a new table with reordered columns.
- **Speeding up the project**
  The goal is to conduct more work in parallel by moving activities to the left. That is, move activities from later phases to earlier phases. Do this for each row.
- **New technology**
  Implementing new technology causes a substantial change in one row and a ripple effect in other rows. For example, different technologies might require different infrastructures or office layouts. You may also change the new activity.
- **Resistance**
  Substantial management or staff resistance to the new E-Business activities may necessitate a change in the pace of implementation. You can do this by stretching out the duration of each phase or by adding phases.
- **Diversions**
  During implementation, you may be diverted into other new areas. This will add work to your rows within a specific phase. Use the table to show the diversion and the impact of a shift and delay due to the diversion.

- **Technology failure**
  When the technology on which you pinned your hopes and dreams has failed, return to the table and redo the technology-related row. Make adjustments to other rows reflecting the impact of the technology change. The table can reveal the impact of technology substitution.

STEP 5: DEFINE THE PROTOTYPE AND PILOT ACTIVITY

Define an initial version of the business activity, and then develop the *prototype,* which is a working version of the system that lacks major databases and files. Functionality of the system is not complete.

You can test the prototype with department staff. You can use a demonstration approach. For example, you can employ an approach of continuous demonstrations. These were conducted for three hours a day, across a month, in Marathon's case. Work can continue to enhance functionality. You can begin to flesh out the new business activity.

When the prototype has been tested and evaluated and is complete, it can be united with the new business activities and tested. This is the *pilot* activity. At the end of the pilot work, the system and activities are together and ready for implementation.

Benefits of this approach are the following:

- The prototype tests the infrastructure and network.
- The prototype provides for tangible feedback from staff.
- The approach reduces the documentation required.
- The business activity has more meaning with the prototype system in place.
- Through direct involvement in the evolution of the prototype, staff and managers begin to feel that they have a stake in the outcome and commit themselves to the task.

Some drawbacks are as follows:

- People may think of the prototype as the production system. This is a common problem with any prototyping approach.
- The initial work may take longer, due to the prototyping and pilot work.

STEP 6: TEST YOUR COMPLETE E-BUSINESS IMPLEMENTATION STRATEGY

Once you have developed the table representing your strategy, use the following steps to test your strategy:

1. Check that the rows are complete.
   Each row represents the major focus of a component by phase. Make sure that these activities are clearly defined and complete.
2. Make sure that the columns are consistent.
   The columns, which show phases, have specific known dependencies. For example, elements of the infrastructure typically must be in place prior to installation of the new activity and system. Looking at the entries in adjacent columns can assist in determining dependencies.
3. Make sure that the phase end products have been listed.
   What end products occur at the end of each phase? Are both the business in general and the activity in particular improved?
4. Look for gaps.
   When there is a blank entry in a row with entries in the surrounding columns of the same row, you have a problem. You will lose momentum because you will have considerable dead time while work progresses in other rows. Consider moving entries around to eliminate such gaps.
5. Test strategy flexibility.
   Mentally project the results of some of the changes covered. These changes affect rows and echo down columns. What happens to your table after these alternations have been carried out? If the number of columns expands, this may mean that the project is becoming more sequential, which is bad if you are trying to meet a deadline.

STEP 7: REVISIT THE COMPARISON TABLES

Carrying out the implementation strategy will produce the benefits defined in the comparison tables. Review the implementation strategy table. Define the benefits and results achieved for the completion of each column. Completion of the work in a column should yield benefits and progress toward the objectives. This is another test of the implementation strategy.

Presenting the strategy as a table will facilitate understanding. To gain support for the strategy, show how the column results support the comparison tables. Managers may pose various questions in the form of, "What happens if . . . ?" Answer these questions in the context of the strategy table. Show what happens when you change and reorder the table entries.

In the management presentation, also show alternatives. Two that you can frequently employ are a go-slow implementation and an accelerated strategy. These can be compared in the tables. They can also be discussed in terms of costs and benefits. Remember that the costs mostly relate to infrastructure and tend to be incurred in the early stages of implementation. The benefits that you present should include those that will be achieved as implementation stages are completed. You want an implementation strategy that yields intermediate results. If the benefits are at the end of implementation and the costs are at the start of implementation, you may have a major problem on your hands. It is important to discuss in rough terms cumulative costs and benefits over time. A graph showing the crossover point when cumulative benefits exceed costs is useful.

Here are some questions to ask. Can you explain your implementation strategy clearly and concisely? Can you explain why you developed an implementation strategy? Have you tested your strategy by assessing the effects of change or crisis in the project on the implementation?

You can also determine if you are getting your point across. Are people using your implementation strategy in their documents, presentations, and conversations? If they are not, you likely have a problem in lack of acceptance. Alternatively, you may not have explained the strategy clearly.

E-BUSINESS EXAMPLES

RICKER CATALOGS

Ricker did not have an E-Business implementation strategy. Instead, they decided to move ahead with implementation planning. That this was a mistake was evident as the plan had to be changed many times at the beginning. Having an E-Business implementation strategy would have saved time.

MARATHON MANUFACTURING

Marathon took the time to define the implementation strategy. They drew up a number of versions of the implementation strategy table that was presented in this chapter. By doing trade-offs, the managers and senior staff better realized what they were getting into during implementation.

ABACUS ENERGY

Abacus required a simple implementation strategy. Using the table they found that a critical issue was to get suppliers involved in the implementation prior to going live. This resolved many potential problems early.

CRAWFORD BANK

Crawford developed several alternative implementation strategies. This allowed them to assess the impact of the E-Business implementation on their normal business. The bank also developed the table and coordinated changes in internal activities along with E-Business ones.

E-BUSINESS LESSONS LEARNED

- **Determine whether each entry in the table is a project or multiple projects.**
  Get rid of multiple projects in the table by splitting them into separate sentences or lists so that each becomes a project.
- **Use the strategy table for multiple business activities.**
  To handle multiple activities, keep the strategy table and create additional rows for the other activities and additional systems. Typically, the infrastructure will apply to all of the activities so that it can remain as one row. The same is true for organization and measurement. This is also a way to evaluate whether your activity groupings were correct.
- **Make sure that the early phases offer little risk in terms of organization and policy.**
  Management will often be reluctant to take major risks early in the project.
- **Watch for too many entries in one column.**
  This may mean that you intend to attempt too much in a short period of time.

WHAT TO DO NEXT

1. Develop alternative implementation strategies using each of the following criteria:

- There is no resource limit
- Very limited resources are available
- You are unable to change the organization during implementation

Define the phases of the implementation for each alternative. This will indicate the differences in focus and emphasis between the alternatives.

2. Determine if your strategy is complete by answering the following questions:

- Are all resources covered by the strategy?
- Does the strategy address interfaces with other projects and activities in the organization?
- Does the strategy allow for contingencies?
- How does the strategy appear from management and business unit perspectives?

3. Test yourself on what you would do if things go wrong or change. How robust is your strategy? For each of the following factors, indicate how you might respond in a way that is consistent with your strategy:

- Management finds a new management concept and asks you to incorporate it within the E-Business project. The concept might involve reward structures, team methods, etc.
- Delays occur in the building of infrastructure involving the first departments in the implementation of the new activity. Should you consider changing the strategy to implement with a different activity group?
- There appear to be delays in building and installing the software. Should you consider implementing other parts of E-Business without all of the automation?
- Management has decided that the project should be speeded up to get the benefits earlier. How would your strategy accommodate such changes?
- Management has decided to deny resources to your project. How will you slow the project down without ending it?
- As you start implementation, you find that an adjoining activity has even greater benefits. What should you do? Should you propose redirecting resources or continue on?

Chapter 12

Action 10: Perform Your E-Business Marketing Activities

INTRODUCTION

GENERAL E-BUSINESS QUESTIONS

There is a great deal to cover in this chapter. E-Business does not market it-self—either internally or externally. Internally, you must get support and participation for the project. Externally, you have to define how you are going to approach the marketing. You also must move the marketing organization to restructure and widen its scope toward E-Business. Some specific items to address are:

- What incentives will be given to salespeople to support the web? If you do not consider this, then they may see the web and E-Business as a threat to their salaries and commissions.
- How will general marketing of the web site be handled? How will you promote the web site in your own marketing and business materials? On what web sites will you post banners and other advertisements?
- What is your marketing approach to get current customers to use the web site? What incentives in the form of discounts such as percentages off and free shipping and handling will you offer?
- What products and services will you offer on the web site initially? What additional items will you offer?
- How will you use the information on web sales and use in terms of analysis?
- What will be the approach to promotions for the web? How will these be coordinated with traditional marketing?

- What new market segments will you be addressing with the web?
- As the number of E-Business transactions expands, how will marketing take advantage of these?

WHY INTERNAL MARKETING IS NEEDED

Even with media and management attention, E-Business implementation does not sell itself. Selling E-Business is more complex than selling standard computer systems, quality management, team management, or many other topics. Here are some reasons for this:

- There is often no natural sponsor or specific department to champion E-Business. It is too new to the organization.
- The business activities cross multiple departments, and each department has its own agenda. The departments may have long-standing hostilities toward each other.
- E-Business involves a large project and requires time and effort; large projects are more difficult to market.
- Numerous points in the project require successful marketing. Failure with any one of these may doom the entire project.
- Improving the organization's and infrastructure's business activity at the same time adds complexity.
- Time-honored policies and roles are difficult to change. Resistance can be both direct and subtle.

Failure to consider marketing is probably one of the leading causes of E-Business implementation failure. People assume that just because management mandates E-Business, everyone will get on board. Marketing includes direct and indirect sales and marketing. When you conduct demonstrations, gather information, review results, and present documents and plans related to E-Business, you are in fact marketing. The care and attention given to marketing, sales, overcoming resistance, and meeting challenges are critical success factors.

In each of the four example companies internal marketing was essential to maintain support and interest in E-Business. E-Business projects often start with a bang and big push. As the detailed work progresses and time passes, interest may flag. Also, new people may join the firm and they may not be aware of the E-Business effort. As a suggestion, you will probably want to visit new managers to inform them of the E-Business project. It is better if they hear about the project from you.

Presenting a new E-Business activity requires careful marketing preparation. When you come up with a new concept, you raise excitement and interest. If you are not careful, you risk raising unreasonable expectations. You also may raise fears of change. Marketing is also important because an E-Business project requires substantial human interaction. Therefore, there is always a risk for mis-understandings. It is not enough to merely present results. Your attitude and your marketing follow-up will strongly impact success.

An example of a marketing misunderstanding occurred at Marathon Manufac-turing. Initially, the E-Business team paid little attention to marketing and the existing sales team, who in turn began to feel left out. Some salespeople felt that they would be terminated and the business moved entirely to an electronic basis. The E-business team could not figure out why marketing did not participate. When the problems finally surfaced, extensive briefings were held on E-Business and the strategies for E-Business with marketing. Marketing then got on board the E-Business train. This shows that what can be a sure success risks becoming a dismal failure.

Consider the situation from management's perspective. A stream of new ideas continues after the date of project approval for E-Business implementation. They receive sales pitches from consulting, technology, and accounting firms. Everyone competes for attention and resources. Winning project approval is only the first step. You must keep on selling the project throughout the project.

Keep a low profile. If you give E-Business and the project too much attention, people can get burned out. Combine a low profile with a constant presence. Seek regular contact with managers and staff to give updates, to bring issues to the surface gradually, and to push for decisions. When people know what is going on in a project, they feel comfortable. They also feel a greater sense of participation.

How Information Gets Misconstrued

Successful marketing means that the correct message on an issue has been conveyed. For example, suppose you hold a project meeting. The meeting goes well. You resolve issues, and the new business activity looks as though it will work. However, most of the time in the meeting is spent in dealing with detailed workflow questions about how the new activity might not address a situation in the same way as the old one. Later, someone asks one attendee how the meeting went. The attendee replies, "We spent most of our time on several areas in which the new approach will not work." This is not the impression you want to create. You have not conveyed the benefits of the new E-Business activity.

Marketing includes damage control. Listen for rumors so that you can correct mistaken assumptions. If you do not correct misimpressions, they will spread. It will become increasingly more difficult to deal with the situation and to confront issues. Damage control takes exponentially greater effort the longer you avoid dealing with it.

The purpose of marketing the E-Business project is not only to gain approval for specific actions, but also to garner enthusiasm to move ahead and to continue support as the project progresses. A long project will have many issues. To handle these, you will need not only sales skills but also an organized approach to internal marketing.

MILESTONES

The major general marketing milestones consist of the following:

- Marketing organization that can handle both traditional business and E-Business
- Identification of potential new markets for E-Business
- Marketing strategies for products, services, and advertisements of the web site
- Approach for salespeople for the web
- Measurement methods for E-Business transaction and activity data
- Approach for promotions on the web

The major end product for the internal marketing of E-Business is support for E-Business implementation. Marketing is the responsibility of the entire project team, including the project leader. Participation and commitment are keys to success. The leader must bring up the subject of marketing and convey information. He or she must define people's roles so that expectations will be realistic. The entire team isn't going to become a sales force, but team members must be aware of the importance and impact of their statements, attitudes, and impressions.

METHODS FOR E-BUSINESS

This section will be divided into two parts: marketing of E-Business internally and addressing the marketing questions raised earlier.

INTERNAL MARKETING OF E-BUSINESS

Step 1: Get Project Approval and Kick Off the E-Business Project

Seek support from middle management to begin an E-Business implementation effort even though upper management has already endorsed the effort. You will work with and seek the support of staff and supervisors in the areas you will investigate. If these people are indifferent or lukewarm, you have little chance of success. Remember that endorsement of a general concept and approval of a project that may be expensive are two entirely different things.

Build up support slowly to catch the interest of departments. As you present the idea to management, stress how E-Business builds on the experience of the current business. Do not claim huge benefits. Do not highlight expensive infrastructure changes or risky organization changes. You do not know enough yet to suggest and support changes of this sort. Your attitude should be, "We're going to make it better, but we don't know yet how much better." Your goal is that the project will begin in a low-key manner. If you are too successful too soon, management may label the project as the salvation of the firm. In such a case, you will set yourself up for failure.

Here are some common obstacles to gaining approval, along with suggestions on how to surmount them:

- The status quo. Defenders of the status quo may raise false reasons for rejecting your project. For example, if people tell you that they are too busy, they may be defensive or afraid of change. Reason with them that no time is a good time, so why not begin the project now? Address what will happen if the current business activity is continued. How fast will deterioration occur? Do not be alarmist, but indicate that changes will be more complex and expensive later. Create a table of deterioration. The table's rows consist of deterioration attributes and its columns of time periods, such as months or years. The entry in the table is the error rate or deteriorated value.
- Other concepts and buzzwords. Two examples are *quality management* and *team management*. Point out that E-Business implementation has its roots in process improvement and work design. Hence, it is stable and can yield more benefits, if properly implemented.
- Other critical business projects. You will want to work out a coordination approach with the leaders of the other projects.
- The timing of other work (the long-range plan, the architecture, the big study). While the subject of the other work is an issue, the argument for going ahead is that these other efforts will not impact E-Business, which is on a fast track for implementation. And by the time the results of the other work are known, E-Business will be either in operation or near to going live and so can use results from these efforts.

Make sure that when people initially approve the effort, they have agreed on what to do next. After the decision and approval, take advantage of the momentum generated by the approval to encourage continued participation. Begin to identify people to be team members.

Step 2: Market the New E-Business Activity

Once you have approval to begin an E-Business implementation project, your marketing efforts will shift to focus on the details of the E-Business activity. The new business activity will receive future support in terms of funding and politics if it is strategically important. So show management how the new E-Business activity supports the vision of the organization.

As in other areas, market E-Business to staff as well as to managers. If the staff members are supportive of the new activity, management will also be supportive. Staff members should be able to explain and articulate the reasons for the new approach. Otherwise, confusion will occur, because you are asking management to endorse a new approach that is unclear.

Your first step in marketing the new business activity is to present the tables discussed in earlier chapters. These comparison tables clearly highlight the differences between the old and new activities. Use the software aids that come with graphics presentation software to prepare presentations more rapidly and effectively. If the staff members participate in this work, they can help you explain the tables to management in an informal presentation.

In the presentation, first give an overview of effort and tasks performed. Call on staff members to explain the details. Then explain how the E-Business activity is better overall for the company than the current activity. Use this same approach to present the new activity to upper management.

Always present alternatives (see Action 7). Alternatives will indicate that you have done your homework. Also, alternatives can offer more palatable versions of the plan. For example, an alternative activity may use no additional funds or involve no organization change. Employ staff members to present the details.

In getting people to accept the E-Business activity and the modified standard one, one of your goals is that the people be willing to end the old way of doing work. One way to encourage this is to pose the question, "How should the old ways of doing work be eliminated?" Ask people to think of all of the actions that are necessary to terminate the activity. Also ask them to verbalize how they will know that the old activity has been eliminated.

You are unlikely to achieve 100% acceptance of the E-Business activity. Focus on the key individuals who work with the new activities, and who will support and reinforce the new activity after you have left. You cannot be around the work forever. When you leave, you want to be able to rely on key people to maintain the E-Business activity.

If you fail to win approval for the new E-Business activity, figure out why. Perhaps, your presentation was too detailed and perceived as overly complex. How were you turned down? Did someone devise a better way? Is there a less expensive version of the new activity?

Success in E-Business approval means that you can move on and generate an implementation strategy. In parallel, you can also develop the plan and budget, as well as a detailed list of initial implementation tasks.

Step 3: Market an Implementation Strategy

The marketing of the implementation strategy is a watershed activity. The previous efforts were tied to selling concepts. Now you will market practical plans. This sets the stage for support needed later during implementation when problems and opportunities arise.

The goal of marketing the implementation strategy is to show how the implementation supports the business vision and will actually install the new activities. The marketing effort links the big picture and detailed implementation tasks.

The audience for this marketing step, as for the other steps, includes both staff and management. Employees who understand the strategy are better able to deal with change in the plan and alterations in direction. The strategy will also show managers how their employees are part of the implementation.

With the surfacing of the implementation strategy, business activities and change are made more visible. This may increase resistance among people who perceive the new approach as a threat. This is yet another reason for convincing people of the validity of the implementation strategy.

If you are successful in presenting the strategy, quickly follow up with the implementation plan. If you wait, you risk loss of momentum and a need to re-introduce the strategy. Following up can also head off misunderstandings or confusion raised in the strategy presentation.

Step 4: Market the Implementation Plan

The goal of marketing the E-Business implementation plan is to stimulate involvement and commitment, rather than to obtain funding. You want people to volunteer to work on the project in order to avoid drafting unwilling participants. Your marketing of the plan should show people their roles and their importance to the project. You can indicate future challenges when you present the plan, but also show organization and an approach that is systematic.

The selling of the implementation plan is the selling of your management style and methods. Most marketing failures occur here. The impact of failure is substantial. Give the marketing of the implementation plan a great deal of attention. Be prepared to answer many questions.

In marketing the plan, present it informally to the people you want on the project. If the wrong people wind up on the team, the project is handicapped. Keep your presentation on a general level. Don't rush in with GANTT and PERT charts. Explain the scope of the implementation, what it means to the company, and the big picture. Then move to the detail of their roles in the project. Explain how these roles contribute to the overall project and why they are important. The closing argument is not an attempt to sign them up, but an opportunity to indicate that you are sensitive to their other commitments. Explain that you will use them part-time when possible. In this way you show respect for their time.

As you move from staff members to managers, the same approach applies. Appeal to their self-interest. Indicate that you are going to be making limited demands on their time. Also, indicate how you will be providing feedback and getting input from them.

The sign of success in marketing the implementation plan lies in the people you have lined up for the implementation. You can get all of the money approved, but if you start with unqualified or junior employees, you are headed for trouble that will be difficult to remedy later.

Step 5: Market an E-Business Budget

In this marketing step you are attempting to show that you know how to manage the budget and are careful with resources. You want people to have confidence in you. Thus, in marketing the budget, you are also marketing your management skills. You will want to ensure that marketing has allowed for their effort in E-Business.

The audience for the budget consists of three groups. The first is the staff in the project team. Staff members must understand the budget and areas of risk so they can help control the budget. Managers who keep their staff in the dark put their project at risk. How can a staff member understand whether a decision is important without some understanding of the budget? The second audience is vendors, suppliers, and customers. Outside firms must agree on the budget as it pertains to them. The third audience is management.

In your presentations, begin by identifying the areas of the budget that are subject to change. Concentrate on areas of risk. Remember that the infrastructure bears the greatest financial risk. Explain contingency plans that you have developed and how you are going to control costs. Do not spend a lot of time on parts of the budget that are large but fixed and known.

If you are asked how you will manage the budget, either directly or indirectly, the following approach may be useful. Highlight which areas of the project are not dependent on unknowns in the business activity and can be funded immediately. This removes them from the critical path. Areas of the budget that hinge on the outcome of the prototype and pilot of the business activity remain pending. This mixed strategy of commitment and control works. Also, by having areas of the budget uncommitted, you can gain power over vendors and push staff to achieve specific objectives.

E-Business implementation budgets can be very large when you add up all of the pieces. Divide the budget by phase so that you target incremental commitment and approval on a phase-by-phase basis. However, indicate that if you are successful, you will be seeking a specific sum overall.

Step 6: Provide a Project Progress and Status Report

In a status report you convey an impression of what is going on in the project. This is an important ongoing part of the marketing effort. The audience for status reports includes not only the direct management for the project but also the managers of your project team members. These people can pull the plug or discourage others. You want them to be supporters, so strive to keep them informed.

When you provide progress information to managers or team members, begin with a general view of the situation. This tends to calm people. Then move to the three topics that are often of the greatest interest. First is the budget. Has there been any change? If not, say so. Second, what events and milestones have occurred? Give a general report, not a blow-by-blow description. The third topic is that of issues. Zoom in on a few issues. To generate interest, build up each one in terms of importance and impact. Then indicate what actions are being taken. If you have identified several alternatives, present them. Stress that you are moving toward a decision on how to deal with a particular issue and that you will consult with them prior to making that decision. Do not ask for input or force decisions on how to solve problems. You are simply providing information.

Give progress reports frequently and informally. Do it on the spur of the moment in hallways if that works for you and those who need the information. Keep a casual tone. The audience feels that they are being kept up-to-date and will not have to ask you for status in more formal ways.

How do you head off negative reaction to status reports? Establish regular updates and progress reports. Go in early or stay late and make the rounds. People will trust you because of your ties with the project and because of the rapport you have established.

If you fail in marketing here, managers will seek you out with questions. If you are proactive, this should not occur. Furthermore, success means that managers will be willing to address your issues based on the relationship you have nurtured.

Step 7: Identify Issues and Resolutions

Issue resolution, including the marketing of solutions, is probably one of the most important aspects of managing an E-Business project. In these projects issues rise rapidly as you address major infrastructure, organization, and technical concerns. Your ability to deal successfully with issues will affect your chances for success in the project.

First, develop alternative solutions, as you did in Action 7 when you created a new activity. What happens if nothing is done about the issue? What happens if you throw money at the issue? Whether or not you present these alternatives to others, this exercise will help you to understand the issue more clearly. Also determine if you have identified and prioritized issues correctly. Have you moved beyond identifying merely symptoms?

The participants in this marketing step are project members, department staff, and management. Participating in issue determination is enjoyable for many people. They begin to feel that they are a part of a team. Dealing with issues also permits them to participate in a limited way without devoting themselves full-time to the project.

Relate issues in terms of their source and nature, grouping them similarly as for business activities. Talk about why an issue has surfaced now. Is it due to urgency? Or is it because it is convenient for someone? What is behind the issue?

Address multiple issues through one or a series of decisions. Resolving issues one at a time through the project is too time consuming. Do not rush to announce solutions. To market solutions successfully, keep management informed as to which issues have been addressed. They can then join the bandwagon and support the resolution. It may be better to have a manager announce a solution. He or she can take credit for it and stand behind the enforcement of the resolution. If you are going to let an issue alone, tell people why you have decided to do this so that they do not see you as indecisive and weak.

How do you define success? Did the resolution work? How do you know it worked? Did people's behavior change? Search for signs that the issue is returning in another form. Success occurs if, and only if, the solution sticks.

Step 8: Get Approval to Expand E-Business

You might think that this is automatic. It is not. People may be stressed out and tired. They want a rest. So you will have to market additional effort. Your last step is to be able to move to other areas. The purpose is to continue the momentum of success to other areas. This does not call for formal marketing. If you have performed well, you will be able to start up the new project with management's informal approval prior to the end of the first project. Begin with a low-cost analysis effort as you have done in previous chapters.

Do not abandon the first departments and business activities that you addressed. Straddle the new and future activities for several months while the first activities shake themselves down. If appropriate, have the first department staff share experiences with the new department staff. Have the first staff demonstrate the differences between the old activity and the new one.

GENERAL MARKETING QUESTIONS

How you address the marketing questions presented earlier depends on your own situation. For each, we will discuss how each of the example firms answered.

- **What new market segments will you be addressing with the web?**
  You have to decide if you are going after new markets and customers or drawing off your existing customer base. You may want to use your customer base to get you started and then go after new customers. The potential problem that you must face is the danger of cannibalizing your existing sales channels and moving people to the web where your profit margins may be less due to promotions.

Ricker Catalogs wanted to attract new customers, but knew that their best shot to get going was through their current customer base. Through catalog promotions, they could support the web site. Marathon Manufacturing decided to target smaller machine shop firms who were not existing customers. While notifying the existing customers of the web site, the target was to get at this market of over 400,000 firms. By contrast, Abacus Energy was targeting their existing suppliers. They only wanted new suppliers if the current ones did not support E-Business. Crawford Bank wanted new customers for applications, but realized that they would also move some customers from their current business.

- **What products and services will you offer on the web site initially? What additional items will you offer?**
Ricker Catalogs decided that they could support all of the items that appeared in their current catalogs. Some products are less likely to be purchased on the web because they are best purchased with in-person contact or extensive pictures. However, they also found that they could attract some products that could be offered on the web that would not be appropriate to the catalogs due to a limited potential market. Also, they found that it cost less to establish a product on the web than in the catalog. Ricker thought that the product mix and range would grow much faster on the web. Marathon decided to offer all of their products to the smaller firms they were targeting. However, they also decided to offer services related to instructions, guidelines, and a bidding wizard for machine shop orders. Abacus had no problem with the services since they wanted to automate the top 10 steps and activities for the purchasing and contracting areas.
Crawford Bank decided to begin with several types of loans (personal unsecured, automobile, etc.). They then planned to expand to leasing and home/apartment loans as well as credit cards. The initial focus was on applications and application processing. They then could expand into servicing and other areas.

- **How will general marketing of the web site be handled? How will you promote the web site in your own marketing and business materials? On what web sites will you post banners and other advertisements?**
Ricker Catalogs decided to market through their catalogs along with limited promotions through other sites. Major creative work was done to establish the look and feel of the web. They contracted with a marketing firm to determine the most appropriate web sites. Abacus had none of these issues. They contacted their suppliers directly and had a project team setup to handle the conversion of suppliers into E-Business. Crawford Bank followed an approach similar to Ricker in that they promoted the site through the branches and on limited web sites.

Marathon Manufacturing had more issues than others. First, they had to get to these firms that had never been reached before. There was no general list. They used the sales force to call on the machine shops in key cities. Then the problem was how to reach the other smaller shops. They used traditional advertising and promotions through paper magazines.

- **What incentives will exist for salespeople to support the web? If you do not consider this, then they may see the web and E-Business as a threat to their salaries and commissions.**
  This was an issue at both Ricker Catalogs and Marathon Manufacturing. Ricker had sales people who arranged for products to be placed in catalogs. These people were tuned into manufacturers who could put goods on the web. It took a great deal of effort and thought to come up with an approach for pricing and commissions.
  Marathon had an existing sales force who called on a fixed number of large firms. They decided that they had to use these people to support smaller firms. The salespeople protested that this did not generate any revenue for them. Marathon then gave them commissions on web sales as well as incentives to distribute material to the smaller shops and sign them up on the web.

- **What is your marketing approach to get current customers to use the web site? What incentives in the form of discounts such as percentages off and free shipping and handling will you offer?**
  All of these steps involve risk. If you adopt an aggressive approach, you could seriously damage your current customer base. E-customers may be more prone to switch to other firms once they get hooked on using the web. Another problem lies in promotions. You cannot offer deep discounts forever. However, once people see it, then they may come to expect it. If they do not get the discounts from you, they may not come back to either the normal business or the E-Business.
  These questions were issues for Ricker, Marathon, and Crawford. Ricker decided to offer specific promotions to catalog customers. Marathon offered discounts for first-time purchasers. Since Crawford Bank offered a commodity product, they had to focus on fast evaluation and service. These did not work. They also had to offer promotional rates.

- **How will you use the information on web sales and use in terms of analysis?**
  This was the biggest challenge for all four firms. None of them really used the available information to full advantage. There was a lack of staff resources. Ricker belatedly did analysis and found that they were overpromoting the web site to customers. Marathon found that the manufacturing demands were changing based on orders from smaller machine shops. This had an impact on manufacturing and costs. Crawford Bank conducted stan-

dard management measurements, but initially failed to analyze how the people accessed and used the web site. Many people visited, but few closed loan applications. They finally made the web site more friendly.

- **What will be the approach to promotions for the web? How will these be coordinated with traditional marketing?**
  Ricker Catalogs had this problem from the start and it continues today. It is very difficult to coordinate promotions. If you push one channel, you can negatively impact another. Marathon had no problem in this regard since they were the first web site in their segment. Crawford Bank faced the problem of differentiating the product in the branches versus the web. They finally decided to favor the web.
- **As the number of E-Business transactions expands, how will marketing take advantage of these?**
  Ricker found that they wanted to expand the range of products on the web. They gradually narrowed their paper catalog business. Marathon Manufacturing decided to offer new products exclusively for smaller firms. Abacus decided to expand on the range of transactions with suppliers. Crawford Bank plunged ahead with more products before they expanded the range of services.

Getting People's Interest and Keeping It

One important general marketing issue in E-Business projects that extend over a long time period is how to keep people's interest. Over the years, several techniques have been used, including the following:

- Take a low profile and give out little information. This approach will work for projects that will last less than six months. With longer projects people will begin to ask questions, so this approach does not work for long-term projects.
- Take a high-profile, headline approach. This technique is one pursued by a number of accounting and consulting firms. It works for awhile, but eventually people want to see tangible results. The pitch and crescendo of the headlines often tie only indirectly to the real results.
- Gain different perspectives by viewing the project from various angles. This approach works for long-term projects. Concentrate on detailed procedures in order to involve people in details of the work. Then shift attention to infrastructure and technology. Next, concentrate on policies. These were not brought up for discussion previously and they open up a new view of the activities. Finally, address organizational issues.

To transition between two perspectives, show the links between them. Use issues as the mechanism to present different perspectives.

Self-Evaluation

Here is a list of questions to help you assess marketing results:

- Do you have regular contact with managers and staff members in which you convey information on the project? Do you have an open line of communications with managers and staff members?
- Is the list of issues being tracked and managed?
- What is the tone of the project team? Are there emotional ups and downs? Is there enthusiasm for the project?

Marketing through Documents

Be careful when using written documentation for marketing. Positive projections and statements have a reputation of returning later to haunt you.

When you document, be factual. Avoid emotional prose. Avoid tacit threats in terms of deadlines. If you want a decision, indicate that you and the team await the decision. Do not describe the dire results that will occur if there is no decision. Document status, issues, and budget in a regular pattern. Establish a format, so that readers will not have to cope with different structures. Use short words and keep your writing dispassionate and to the point.

Marketing through Presentations and Meetings

This is a favorite way of marketing. When you present in person, you can express passion and emotion. There is give and take. You can elicit commitment. Another advantage is that, while the multitudes are using Internet and electronic mail, you are using personal contact. Keep the electronic mail in the background as support.

Decisions

Major decisions need to be made by management. However, routine decisions should involve staff members and managers. Involvement in decisions increases commitment. Your objective is not quick decisions. Decisions are a matter of timing and circumstance.

When a decision needs to be made, first lay the groundwork. Present the issue. Review it several times, along with alternatives. Communicate with those who are affected by the decision as well as those involved in carrying out the decision. Here are some strategies to pursue next:

- *Break large decisions into smaller chunks.* Smaller decisions can be addressed at lower organization levels.

- *Do nothing.* Err on the side of inaction. Being too much of an activist and pushing for decisions may alienate management.
- *Don't force an issue.* If you force an issue, you will likely lose support for a decision.
- *Consider marketing to the people around the decision maker.* Attempt to achieve consensus. Watch for signs that you have antagonized someone. Follow up and see if a little more effort can win over this person.

When a significant decision goes against you, this can affect the momentum of the E-Business. People begin to question management's desire to continue the project and management's commitment to E-Business. That is why decision making should be your highest priority.

Once you have a decision, take action immediately. Have an implementation approach ready for launching when the decision is made. If you delay, the force behind the decision abates. People wonder, "If it was so important, why hasn't there been any action?"

A number of suggestions have been provided for each marketing activity. Overall, ask along the way if you are learning the lessons from the previous activities. Are you getting better? Or are you making the same types of mistakes? If management has changed, has your marketing approach been modified?

E-BUSINESS EXAMPLES

RICKER CATALOGS

For Ricker, marketing in both areas was a major effort. The marketing staff were not familiar with the web and E-Business. This required a learning curve. Contacting suppliers to put products on the web was another challenge to marketing.

Internally, constant marketing was required. Ricker had substantial middle management turnover. This meant that new managers had to be brought up to speed almost every two weeks. What made this more important was that these managers oversaw parts of the E-Business activities and staff.

MARATHON MANUFACTURING

Marathon required little general internal marketing except in manufacturing and marketing. In manufacturing, the manufacturing managers did not realize that orders from smaller firms on the web required a rapid response. Smaller firms ordered different machines than larger firms. In the marketing area, the resistance of the sales staff had to be overcome. This effort required more than incentives and commissions.

ABACUS ENERGY

Abacus had almost no internal marketing to do. Their problem was in pushing E-Business to their suppliers. This was a challenge due to the suppliers' lack of familiarity with e-commerce and their long relationships with Abacus.

CRAWFORD BANK

Crawford Bank had internal resistance along the lines discussed in the chapter. People felt that the web site was too impersonal and that you really needed personal contact when lending. This was a challenge to overcome. Externally, the problem was how to identify and reach the right market segments for the loans.

E-BUSINESS LESSONS LEARNED

- **Involve people in minor issues and questions.**
  Examples are the location and number of cable outlets for workstations, status and other codes for the E-Business activity, and policy and procedure statements. This will help you develop a sense of what people are thinking.
- **Continue to spend time with the staff performing the steps in the current business activity.**
  This time helps to maintain rapport and pays off in both support and reduced training later.
- **Sell change and stability at the same time.**
  Make people comfortable by linking to the past. Excite them by linking to the future.
- **Line up support at each level.**
  While implementing E-Business, you cannot afford to rely on only one person. Seek broad-based support at different levels. While this requires a major initial effort, the approach pays dividends later in reduced effort.
- **Use the trial balloon approach for issues to reach decisions.**
  Casually float proposed solutions of issues to generate reaction.
- **Keep a marketing log of contacts and results.**
  Consider building a marketing log for a week. Each day write down what you have done from a marketing perspective. If several days go by with no entries, set aside time for marketing in the near future.

- **Follow up after decisions are made.**
  Make yourself available, both before and after decisions are made. Ask those involved if there is anything else they would like to see performed in the project.

WHAT TO DO NEXT

1. For each marketing step identified in the chapter, list two or three people who are critical to the decision making. Then look across the lists and find the two or three people whose names appear most frequently. This helps define your marketing audience over time.

2. For each decision area identify two reasons people could oppose a decision. Determine what factors could be behind each reason.

Decision: _____

Reasons:

1. _____

2. _____

Factors behind reasons:

1. _____

2. _____

3. Maintain a list of the managers in Action Item 1 with whom you have talked. Include the following information:

Business Activity

Manager	Date	Subject	Comment

Analyze these contacts over time and identify any gaps where you have not maintained contact.

Part V

Implement Your E-Business

Chapter 13

Action 11: Plan Your E-Business Implementation

INTRODUCTION

E-Business implementation planning differs from traditional planning in the following ways:

- Scope. The scope of the E-Business project is wider, involving organization, systems, business activities, procedures, and changes between organizations. It also will indirectly involve customers and/or suppliers.
- Politics. E-Business implementation projects are politically more sensitive and volatile than a traditional project. You must be careful what words you choose in the project plan and in presentations. It is in this action that some people finally realize E-Business will really be implemented.
- Coordination. E-Business implementation requires more people, which involves more coordination because there is a higher likelihood of confusion in the roles and responsibilities of people and organizations. Also, many people have never carried out E-Business implementation and will need proper guidance.
- Subprojects. Many traditional projects are set up as a single project. E-Business projects are different. They usually involve a number of interdependent subprojects that can be done in parallel.

The differences mentioned explain why some E-Business projects fail in implementation when they are planned and managed like traditional projects. To be successful, you must understand these differences and their eventual impact on the E-Business project.

If you put the above bulleted items together, you can see that you want to stress coordination and integration. We feel that this is so important that you should have separate subprojects for all of the major integration efforts in the E-Business project. Some examples are:

- Interfaces between legacy and new e-commerce software
- Interfaces between marketing and IT for requirements and implementation
- Management interfaces for coordination
- Organization change that must be undertaken in parallel with IT and business activity work

THE FIRST VERSUS LATER E-BUSINESS IMPLEMENTATION PROJECTS

E-Business projects also differ from other projects in that they often involves multiple, successive projects. Just as soon as you have the initial E-Business site running, there will be demands for changes to respond to marketing, competition and the internal organization. Standard projects that have a specific scope tend to fall into patterns. Each E-Business project tends to be different—even in the same organization. Methods, tools, and lessons learned in one project can be applied to the next, but the actual method and political and organizational situation will be different. Therefore, it is important that you view this project as just the first of several and not a one time effort.

IMPLEMENTATION RESISTANCE

It would be nice to assume that everyone is on board the E-Business train. With the exception of Marathon Manufacturing there was passive and even some active resistance. Experience shows that you need to be ready for this. It will even occur in spite of top management support. Here are three common lines of resistance.

- E-Business will undermine our current business and customers. If not carried out carefully in this and the next action, this could become true. Therefore, it must be carefully watched.
- There are so many hidden costs and issues that E-Business should be attempted over a much longer time. The implementation plan must be comprehensive.
- The technology and knowledge of E-Business is limited because it is new. Why not wait? The same argument was made with the automobile, the telephone, and the computer.

You can also run into problems in planning. Here are some problems we have had to address:

- Lack of project organization. No detailed implementation plan exists.
- The plan is incomplete, missing infrastructure or other tasks.
- The plan may reveal a hidden agenda. For example, downsizing was not proposed in E-Business, yet it is evident in the task lists.
- The project manager and team lack experience, so the plan may be incorrect or incomplete.

Heavy, successful initial activity can mitigate such attacks before they become serious. Before proceeding further, define known and potential project enemies. Target those people as possible members of the project team. This entails risk. These people can create issues within the team and may attempt to undermine the project with their management. However, excluding them offers greater risk.

The implementation plan has several objectives. First, it is obviously the blueprint for carrying out changes. Second, it helps to reinforce E-Business by supporting the strategy. Third, the planning effort is a major way to engender support and gather momentum for implementation. The fourth objective is political. In your plan, line up project team members by working with their management. This shows management that the E-Business effort is serious. These objectives are in addition to the standard reasons for developing a plan—to track work and progress.

The scope of the plan is comprehensive and includes all parts of implementation, including changes relating to infrastructure and organizational change tasks. The more detailed and encompassing the plan, the better.

MILESTONES

End products of E-Business depend on the E-Business strategy and the nature of the business and its customers and suppliers. Specific end products will be different in content, but can follow a general structure. This is called a template. It consists of standard high-level tasks, assignment of general resources to these tasks, and dependencies. The template does not include the detailed tasks, specific resources, and durations and times for the tasks. These are set by the plan. You will want to define the template first. Then you can use this template later. You will use multiple templates for each major subproject. You want to use subprojects to highlight risk and to ensure accountability.

Here are some of the subprojects that we have employed:

- Project planning
- Marketing
    — Structure of the marketing group and staffing
    — Assessment of the company products for E-Business

- — Identification of potential promotions, discounts, and pricing
- — Marketing strategies for both traditional and E-Business areas
- — Implementation of ongoing monitoring of competition and the web
- Accounting
  - — Analysis of accounting activities and policies
  - — Implementation of changes in activities and policies
  - — Establishment of credit card transaction processing
  - — Agreements for authorization and processing of credit card transactions
- Infrastructure
  - — Office layout and department locations
  - — E-commerce utility software
  - — E-commerce hardware
  - — Networking
  - — Computer security
- Systems
  - — E-commerce application software
  - — Identification and negotiation for contractors and consultants
  - — Internal IT staffing reviews and prioritization
  - — Prototype of the system
  - — Pilot project of the new business activity and system
  - — Infrastructure (nontechnology)
  - — Technology infrastructure
  - — System development
  - — Integration and testing
  - — Data conversion
  - — Business activity policies, procedures, and training materials
  - — Organization change
  - — Training and installation
  - — Cut-over to new business activity
- Interfaces between legacy and new e-commerce software
- Interfaces between marketing and other departments
- Management reporting and measurement
  - — Measurement methods and tools
  - — Establishment of management reporting and analysis

The content and detailed activities behind these tasks are covered in detail in Action 12. Attention is on the plan in this chapter, and phases of the project are defined.

Since implementation planning is a planning activity, you might be tempted to do it yourself, but this is not a good idea. Up to a point, the more participants you have the better. You will give them task areas and they will supply detailed tasks and schedules. This will help in them understanding the overall plan and committing to dates.

METHODS FOR E-BUSINESS

STEP 1: DEFINE THE PHASES OF WORK

Since E-Business implementation will take some time to complete, partitioning the project into phases is a natural first step. Each phase should include demonstrable and tangible results. If you were to line up what you have achieved at the end of each phase, you should be able to discern a cumulative momentum of change.

Breaking down the project into several phases (e.g., activity and group) is one approach, but there are other ways. Instead of two phases, you could associate a phase with each central business activity. This is preferable to partitioning by departments. Departmental partition tends to reinforce the separation of departments. This is exactly the opposite of what you want—more collaboration and cooperation. In E-Business you may want to break up implementation into the following phases:

- Infrastructure, hardware, network, and software established
- Marketing campaigns initiated
- Software completed and tested
- Key E-Business activities in place

However, you can also define phases based on infrastructure, geographical location, and technology.

Putting these comments together you can see a pattern. Have separate parts for each business activity as well as for the E-Business infrastructure, systems, and technology. Remember to keep integration and interfaces as a separate entity to support tracking. There are several benefits to breaking up the project in this way. First, you and others get more visibility into the project. Second, you can ensure that there is greater accountability. Third, having many smaller plans of reasonable size makes the planning method and the project easier to manage.

STEP 2: DETERMINE MAJOR MILESTONES AND E-BUSINESS SUBPROJECTS

Each phase must be divided. Approach this in a top–down manner to ensure consistency across the plan and to make sure each subproject fits with the implementation strategy. Rather than deal with tasks, define key milestones in each phase. Once you achieve the set of milestones, you will know that the phase is complete.

Using the approach discussed earlier, you could define sample milestones as follows:

- Project plan
- Marketing
  - New E-Business marketing organization
  - Marketing business activities
  - Competition assessment methods
  - Marketing strategies
  - Marketing campaign initiated
- Accounting
  - Credit card agreements
  - Credit card processing in place
- Prototype
  - Definition of prototype completed
  - Initial prototype ready
  - Stable prototype achieved
  - Completed prototype
- Pilot project
  - Pilot project defined
  - First pilot project results obtained
  - Second version of pilot project ready
  - Completion of pilot project
- Infrastructure
  - Completion of building plans
  - Approval of building permits and approvals
  - Completion of subsystems
  - Building inspections of specific subsystems (e.g., electrical, water)
  - Completion
- Technology
  - System and network architecture defined
  - Cabling and data communications completed
  - Installation of hardware and system software completed
  - Architecture is application ready
- System development
  - Design completed
  - Development completed
- Software packages
  - Identify the software packages
  - Evaluation and selection
  - Installation and setup

- Consulting support
  — Identify the potential consultants
  — Evaluation and selection
  — Contract negotiation
- Integration and testing
  — Test scripts completed
  — Integration starts
  — First systems test
  — Integration completed
  — Testing completed
- Data conversion
  — Definition of conversion approach completed
  — Completion of programs for automated conversion
  — Manual conversion starts
  — Conversion ended
- Business activity policies, procedures, and training materials
  — Outlines of all materials completed
  — Policies reviewed and approved
  — Training materials tested and completed
  — Procedures completed
- Organization change
  — New organization and details approved
  — Reassignment of staff and managers completed
- Training and installation
  — Training started
  — Training completed
  — Systems installed and tested with converted data
  — System ready for cut-over
- Cut-over
  — Cut-over of systems
  — Cut-over of business activity
  — Cut-over of policies

For each of these you define high-level tasks that lead up to these milestones. The result is an E-Business implementation template.

STEP 3: DEFINE THE DETAIL WITHIN A SUBPROJECT

You can now define a subproject and the detailed schedule for a subproject. Start by defining the next level of detail within a subproject. Assign each area to a team member. Your political goal is to involve team members in the planning approach, and by so doing, to build commitment. Provide the team members with an example to indicate the level of detail needed for the plan. Before they start, hold a group meeting; attempt to get consensus on a list of resources. This will serve as the resource pool for all of their projects.

Each team member should now develop detailed tasks and task dependencies and assign specific resources to tasks. This is the "what" and "who." The duration and dates ("when" and "how long") will come later. This omission is intentional. Doing it all at one time is too much and team members may manipulate tasks to meet schedule dates.

Assign resources carefully. Assume that the tasks are in an outline form in which lower-level tasks fall under and within summary tasks. Assign the person who is responsible for the task area to the summary task. Assign only critical resources to the detailed tasks under the summary. Consider people, facilities, and equipment. Do not include resources that are overhead or that are plentiful. Also exclude common support roles. By doing so, you will have a small, manageable set of resources for each task.

Early involvement of the team not only heads off future problems, but also gives team members an opportunity to see the overall plan. If you wait to involve people during implementation, it may be too late. With no active participation in the project planning, team members will need you to spend more time explaining the project. Another problem that may arise is that staff members may revisit changes, which is time-consuming and unproductive.

Assume that you know the "what" and "how." To review these with the team members, begin by showing them how their tasks roll up into the overall plan. You are now acting as a schedule integrator. Here are actions to take before you ask them to assign dates and durations:

- If the schedule is on project management software program, run a "what if . . . ?" analysis to schedule copies of the team members' work. Suggest possible changes. Do not force them to make the changes; these are only suggestions.
- Describe the difference between duration and effort. For example, if a task requires four days and a person is assigned half-time, the duration is four days while the effort is two days.
- Make sure that team members do not force dates when they fill in the details. Suggest that each task be scheduled as soon as possible to provide flexibility.
- Have team members update their own schedules. Track the changes.

STEP 4: DEVELOP SCHEDULES FOR THE ENTIRE ACTIVITY GROUP

When addressing activity groups, you may find that some organizations support or perform multiple activities in a group. The following are suggestions in handling development of the schedules in the group:

- Teams and departments concerned with one business activity can be treated as suggested above.
- Support teams that support multiple business activities will need separate infrastructure schedules for all activities in the group. Place summary tasks in the appropriate schedule.
- Begin with the schedule for critical activities in the activity group. Then split out the work for the infrastructure and technology that crosses activities. Develop these schedules next for all other activities in the group. Then return to the activities with the summary tasks.

STEP 5: ANALYZE THE PROJECT PLAN

Work in analyzing the project plan includes the following:

- Add milestones where management decisions will occur. This highlights the subjects that management will review. It will be important and useful during the review of the schedule. Assign management as the resource to these milestones so that you can later pull them out and highlight them.
- Make sure that you have regular milestones for each subproject so that you can track progress.
- Ensure that you have links between projects. This is important since you will be the one coordinating between the subprojects.
- Look for dependencies you can eliminate by parallel effort. Try to have consistency in dependencies so that most links are between tasks and milestones at the same level of the outline. This will make managing the schedule easier.
- Find all tasks assigned to a specific resource type. Do you require this resource in each of these tasks? Is the resource crucial to the task? Review tasks with the largest number of different resources and determine why they are assigned so many resources.
- Review the wording of the tasks. Is the wording of a compound task complex? If so, consider dividing up the task.
- Edit the schedule so only summary-level tasks are given. Assuming inevitable changes to the schedule, are these summary tasks complete and stable?
- Build a glossary of terms. Include the definition of each resource and its role, as well as what constitutes each major milestone.

Extract separate subprojects for each involved organization that is providing resources. Answering the following questions may prompt further changes:

- Is too much being demanded of an organization over too short a period? If so, review each resource assignment and the extent of involvement in each task.

- Does the schedule contain gaps when an organization is not involved? If so, is this politically acceptable? You may want to keep people involved and up-to-date. If you have heavy use intermingled with periods of dead times for some resources, you may have a problem. If people leave the project and are reassigned, how will you get them back? Loss of contact means increased overhead as you bring them up-to-date.

Plan ahead for schedule changes. Use the outlining feature of the software. You want the high level of the schedule to be the same despite changes. Detailed tasks can vary. This will allow analysis and comparison of the planned or baseline, the actual, and the "what if . . . ?" schedules.

Once the analysis is completed, return the schedule parts back to the team members for revision. Volunteer to help if they request this. Attempt to get a final plan after two or three iterations of changes.

Assessing Risk

To mitigate risk, develop a list of assumptions related to the schedule. These assumptions often relate to the conditions of the facilities, the technology learning curve, the condition of existing data and use in conversion, and training time required. Project risk often occurs in integration. Consider setting up a special task group within the team to develop the integration tasks. You may wish to have someone on the team serve as the integration czar.

Individual tasks in subprojects will involve risk. Recall that we treat risks by associating each risky task with an issue. Therefore, you should have a long list of detailed issues that map many to many with the tasks. For each risky task there will be one or more issues. Each issue will apply to one or more tasks.

Evaluating the Project Plan in Terms of Your E-Business Strategy

Here are some tests to apply to your E-Business strategy:

- Does the implementation plan yield tangible benefits at each stage that support the E-Business strategy?
- Are the stages consistent and cumulative or are they separate and unconnected? The phases of work in the plan should show that as time progresses, you are moving closer to fulfilling the E-Business strategy.
- Are political, technical, and managerial issues spread out during implementation? Spread the issues and risk associated with the strategy throughout the implementation plan. If they occur too early, there may not be enough available information.

STEP 6: COMPLETE THE OTHER PARTS OF THE PROJECT PLAN

Indicate the method of change control in the implementation plan. Tracking and relating issues to the activities and to the tasks is another important part of the planning approach. While these points are in the strategy, it is the plan that tends to be used by more people.

Of particular interest here is which milestones you will evaluate in detail in terms of content. It would be nice to say that you will evaluate all milestones. However, this is unrealistic given the scope of the E-Business project. The suggestion here is to concentrate on areas of risk. These include marketing, accounting for credit cards, infrastructure, the application software for E-Business, and integration between E-Business and traditional activities.

STEP 7: ANALYZE SCHEDULES FOR THE E-BUSINESS ACTIVITY GROUP

Assemble the individual schedules in a single project for analysis. Along with the specific suggestions mentioned earlier, do the following:

- Filter all schedules by the same resource area. Then combine the filtered schedules. This will provide an overall pattern of resource demand by area.
- Ensure that the schedules for the infrastructure and technology fit the work on the activities.
- Proceed by adding coordination, testing, and integration tasks between activities in the group. Put these in a separate schedule because there is a risk that they will be buried in the individual schedules.

Figure out how coordination will be performed across the activity group. How will changes and schedule slippage be implemented?

STEP 8: SHOW YOUR PROJECT PLAN TO THE TEAM AND MANAGEMENT

To complete your plan, show it to the team. This is part of building team spirit and enables the team to visualize the big picture. To give the members a feeling of progress, show both the first and the last versions of the plan. You can then comment on the positive changes to the plan.

In this team meeting, ask the team members what they got out of the effort. If they were to do it again, what would they do differently? This is the first conscious effort to establish a pattern of recording lessons learned. At the end of the meeting, indicate that you are going to make any last changes needed and present the plan to each of the team member's managers with the team member present.

Management Contact and Presentations

During the work on the plan, stay in regular contact with management. Give out the list of phases, subprojects, and milestones. Provide copies of the resource pool. At this stage, you are seeking passive understanding. You have not provided the dates, but if you wait to give out the entire plan, you may encounter more problems. Management will be more likely to revisit task definitions.

When you present the initial materials, ask who should review the work. This will yield a list of potential reviewers who are highly regarded by management. Prior to reviewing the plan with the team, provide management with a summary high-level schedule for additional feedback. You are seeking major concerns, not approval. If concern is expressed about the schedule, indicate that you will review it and see what can be done.

You will be making presentations to each manager before you present the plan to the highest level managers. The presentations are intended to resolve as many questions at as low a level as possible.

Materials needed for the presentations include the summary schedule, the detailed schedule, the implementation strategy, the detailed subschedule that uses the managers' resources, and a summary of resource use over time.

The sequence of presentations is important. To gain experience and sharpen your skills, first schedule some of the groups that are only peripherally involved. Later, make the presentation to the main owner of the key activity. After this presentation, make any revisions and return to the first groups. Their support will be key. Next, go on to other groups. You can finish with departments with lesser roles.

Tell your presentation audience members that they can bring others. Begin the presentation with the management summary. Show the detailed schedule next to indicate how much work went into the effort. Drill down to the subschedule. The person on the team responsible for the tasks should describe them and make any necessary comments. After the presentation make any changes and go on to the next department.

If you have been successful in the individual presentations, the main management presentation will be easier. For the major presentation, start with the summary of the tasks. Then present the management decision points. Emphasize management's role and possible decisions. If this goes well, you can show additional detail for the first set of tasks in Phase 1. If possible, the team should attend the presentation.

Review for the Activity Group

The review for the activity group after analysis can begin with the organizations that support multiple activities in the group. Then you can move onto the individual activities, starting with the most critical business activities. A variety of dependencies exists between business activities in the group, so consider the most closely related activities.

Setting the Baseline

After the management presentations have been completed, you are prepared to set the baseline plan. In most project management systems, multiple schedule dates are allowed for each task. There is the planned schedule that corresponds to the baseline, an actual schedule where you will record progress, and the working dates. You will get the working dates each time you load and access the software for scheduled dates. This adds up to three sets of start and finish dates for each task that you can compare and analyze.

E-BUSINESS EXAMPLES

RICKER CATALOGS

The E-Business implementation plan was defined by consultants. They followed the steps above and involved many employees in the creation and review of the plan. This made management review much easier. There were 10 subprojects. Each subproject had an average of 300 tasks. So you can see the need for subprojects—putting everything in one plan will ensure that nothing is easily found. Risky tasks will not be given attention.

MARATHON MANUFACTURING

Marathon first had each department manager develop a plan. This was a disaster. None of the plans were compatible with each other. The level of detail was uneven. To save the effort, the individual plans were collected and the approach in this chapter was employed, using the plans as a starting point.

ABACUS ENERGY

Abacus created one overall implementation for E-Business for their suppliers. It quickly grew unwieldy. They also did not follow a template. The plan had been developed by a few people without involving many people. People neither bought into the plan or understand what they had to do. Many meetings were required to redefine the tasks. The one large plan had to be later broken up into subprojects.

CRAWFORD BANK

Crawford had carried out many projects before so that there were project management skills in place. However, they had never done an E-Business implementation. The result was that they missed many tasks. Unfortunately, the missing areas were uncovered one by one. Another problem was that they gave too much attention to the critical path—like good, traditional project management types. However, in E-Business it is the risky tasks that undermine the schedule and the project. After all, who is to say that the risky tasks are on the critical path? Our experience in E-Business indicates that they are not.

E-BUSINESS LESSONS LEARNED

- **In the implementation plan, factor in the learning curve of the team working together.**
  Many managers just assume that people will work out any interpersonal problems. In projects with lower visibility and less pressure, this may be true, but not here. Allow more time for initial tasks.
- **Include in the scheduling any delays in management review and approval.**
  Do not assume that management will instantly act or react. Anticipate a delay in review and approval. Plan for other tasks so that the project is not delayed.
- **Factor in the learning curve with the new technology.**
  Allow for the necessary effort on the part of the employees to gain limited proficiency. This tends to be overlooked in the presence of schedule pressures.
- **Remember to include system interfaces between current systems and new systems that will support the new activity.**
  Here is a technical area that is often poorly estimated and managed.
- **Apply standard project management methods.**
  Employ any standard techniques that your organization employs that are consistent with the methods described in this chapter. Deviations make the project stand out and attract attention.

- **Do not stress the uniqueness of the E-Business project to the team or to others.**
  Keep a low profile with the plan. Make sure it resembles other plans on the surface. The more that your plan is perceived as different, the more likely that the approval time will be delayed.

WHAT TO DO NEXT

1. Write a task list for implementation, including the highest level 140–150 tasks for each subproject. Use indentation to show how detail tasks fall under summary tasks. Label the tasks with task numbers. Add dependencies to the list. You can enter this information into a project management software package or a spreadsheet.

2. Create a resource list for the implementation plan. Make sure that your list is complete in terms of involvement by business areas and contractors for infrastructure work. If you do not know a specific contractor or department, indicate the function (e.g., cable installer). Provide an abbreviation for each resource. This information can be pasted into the project management software.

3. Assign the resources to the highest level tasks in the list you created in Action Item 1. Assign someone to be responsible and to do the work. Plug in no more than three types of resources for each task in terms of work. Note that you can distinguish between responsibility and work because an internal manager might oversee the work of several contractors. If your project management software does not allow for a separate field for responsibility, use a comment text field for this.

4. For each summary or major task area, identify the milestone. The milestone is what you will have achieved if all of the tasks under the summary task are completed. Examples are the system is in place, the building is ready, or the activity is ready for implementation. Write down how you would know if the milestone were achieved.

5. Identify 10 or more tasks in which you perceive risk. Rank each of these tasks in terms of likelihood of task slippage and impact on the schedule if slippage occurs. (This is likelihood of loss and exposure.) Use a scale of 1 to 5 (1 is low; 5

is high). Note that you can have low likelihood, but high impact. Create another column to show the result of multiplying the likelihood by the impact: this is the risk in simple terms. Also identify the estimated duration of each task. Keep this table handy and update it often. It tells you where the real risk lies in the project. People often pay too much attention to likelihood and duration and not enough to impact.

Tasks	Likelihood	Impact	Risk	Duration

6. For each of the high-risk tasks you have identified above, chart whether the risk can be reduced by the items listed below. Use a scale of 1 to 5 (1 is not applicable; 5 is very useful). You can add more alternative actions. The ones listed below are common responses. Adding resources can include not only people, but also tools, methods, and facilities. Breaking up the task means that you will divide the work into smaller tasks to micromanage the work, thereby lowering risk. Extending the time would be useful if the extension would reduce exposure. "Improve Quality" means selecting better people or resources.

Tasks	Add Resources	Break Up Task	Extend Time	Improve Quality

7. Make a list of implementation issues that you think you will face in the project. Now review the task list and find the tasks that are associated with each issue. Next, for each task or issue, determine the underlying issue that leads to the risk. This exercise will help you validate the tasks and risks as well as the issues associated with implementation.

8. Define several alternative approaches for dividing the implementation into subprojects. One is to divide it by areas of risk (highlighting integration). Another is to divide it by functional area (technology, software, business activity, etc.). What are the advantages of each of these?

Chapter 14

Action 12: Execute Your E-Business Implementation

INTRODUCTION

The implementation of the new business activities can be divided into two parts. There is the implementation of the infrastructure, software, hardware, and E-business and standard business activities themselves. Then there is the implementation of changes in the marketing and other organizations. Both will be considered.

E-BUSINESS ACTIVITY AND SYSTEM IMPLEMENTATION

The first part ends with the pilot project implementation of the new E-Business activities and support structure. It includes infrastructure and systems work. Then you will be completing the development, installing the new activities into the organization, and cutting off the old activities. This effort also entails conversion, integration, testing, training, cut-over, and organization change. Your objective is to have the new E-Business activities supported by infrastructure, systems, and organization at the conclusion of implementation.

You have a strategy and plan for implementation. One of your first steps is to develop prototype systems to support the new business activities. Why develop a prototype? Because you want to see how potential customers or suppliers might navigate the site. At one time the concept of a prototype implied a model of the system. The prototype provided an idea of the user interface and data elements, but it could not process information. Today, many prototypes still lack interfaces to other systems, substantial logic, reports, and any batch processing. This sounds like much of the system is eliminated from the prototype. However, the prototype does encompass the user interface and the ability to enter, update, and view transactions.

Successive prototypes will be developed and delivered for review every few weeks. The prototype will never go into production. This is the basis for system development and business activity evaluation later during the pilot project. The prototype effort will end when the number of changes diminishes. Defer some changes until after the pilot project effort.

The pilot project takes the prototype, builds the new activities with the systems, and tests the new activities and systems together. The production approach goes beyond both. Production encompasses integration of all components, testing, conversion, and widespread use. You will use employees and others to test the E-Business prototype in real transactions for ease of use, navigation, understandability, completeness, and other criteria.

What would happen if you combined the pilot project and prototype into a single prototype? You would work simultaneously on changes to the system and to the activity. This makes it more difficult to finish the system and the activity concurrently in terms of getting a stable product. Also, it is difficult for staff members who do not develop systems to attempt to work with both at the same time.

There are dangers to prototyping. Prototyping can be very seductive. You get positive feedback. Changes are being made. There is visible progress.

- Departments such as marketing can push to use the prototype in production. This can be discouraged at the outset and throughout implementation.
- Requirements can waver back and forth. In such a case, schedule a meeting to discuss different approaches and to have decisions made.
- Prototyping and pilot projecting can bring to the surface a number of issues beyond procedures and systems, such as faulty or incomplete policies or misconceptions about the business activity and variations in practices.

ORGANIZATION AND MARKETING CHANGES

There are several areas of the organization that are affected by E-Business implementation.

- Marketing must be changed to analyze E-Business data and to develop promotions, advertising, and discounts for E-Business.
- Customer service may need to be adjusted to handle E-Business customers.
- Accounting may have to have additional functions.
- Inventory levels may have to be changed to ensure that there are fewer backorders.
- Fulfillment and shipping may have to be changed to speed up customer delivery.

PURPOSE AND SCOPE

You are seeking to build a prototype system and new business activities and then test or pilot project these together. Doing this will achieve the following goals:

- Validate that the new activities and systems work together and can replace the old activities and system
- Demonstrate and improve the prototype so as to improve the new activities and market and gain support for the new activities
- Validate the benefits and costs you identified earlier

The scope of the action includes the activities, system, and the organization. This is the first opportunity for the current organization to interface and work with the new activity and system. It will provide information on what organization changes are possible.

On the organization side you have to implement changes in the marketing organization toward E-Business as was discussed earlier. You will also make other changes in departments. For example, you may have to implement credit card processing—something that the organization has never done.

MILESTONES

There are a significant number of milestones that support two basic end products. One end product is the set of working E-Business activities. This has implications for infrastructure, networks, hardware, software, organization, and staffing. But you can have a functional site that is not supported by marketing and other business activities. Thus, the other major end product consists of the marketing, accounting, fulfillment, and other department activities to support E-Business.

Some specific milestones are:

- Prototype of the system working
- Pilot test of the new E-Business activities and system successful
- Infrastructure and architecture changes and upgrades are tested and working
- Interfaces with current systems implemented
- Data conversion and setup for E-Business completed
- Documentation and training completed
- New E-Business activities implemented
- Advertising for the new E-Business underway
- Promotions established
- Initial products and services are loaded on the web; information is set up on the web

- Marketing restructured to support E-Business
- Accounting and other departments have changed to accommodate E-Business

METHODS FOR E-BUSINESS

As in the Introduction, the methods will be divided into two parts.

E-BUSINESS ACTIVITY IMPLEMENTATION

Step 1: Design and Develop the Prototype

Internet and web systems are systems based on network, microcomputer, and database technology. Internet systems are a special case of client–server systems where the web browser acts as the client. The software resides on the server. For a prototype you might have a skeleton of the server and a fairly complete client. This was the case in the examples. In an E-Business application you want to stabilize the user interface and nail down the navigation that customers or suppliers would employ.

You should now be able to develop the initial prototype without a great deal of additional input and data collection. Begin with the workflow. Associate a status code with each workflow step. Identify which step triggers a status code change. Workflow is extremely important because it provides the guide to navigate through the screens. Note also that a key advantage of the new business activity will be the workflow navigation.

Define the databases for the prototype. These consist of core databases and support tables or databases (e.g., zip code lookup). In a recent E-Business insurance prototype the core databases included customer, policy, property, and insurance lines. Support tables included letter generation, policy amount from a code, flood zone area, and status codes.

Next define the characteristics of the data elements for the prototype. Decide on the approach to accessing the databases. This is the access or index key(s). The selection of the data elements for the keys will affect the performance and design of the system. The key is the basis for the sorting of the records in the database. Picking the wrong key can lead to disaster because the system response time might be excessive, or you might not be able to reach the right information.

With the menus, workflow, and data elements defined, define the screens and the GUI (graphic user interface). Try to use tools that can be reused in production such as Java or Active Server Pages (ASPs).

In selecting the tools and methods, employ the following criteria:

- Quality and completeness of the development library
- Additional software tools that you can purchase to reduce development time
- Need of hiring an experienced programmer who can teach your IT staff lessons learned
- Completeness of tools to reduce development time
- Flexibility to make changes quickly in the prototype
- Reusability of the computer code to move to a production system

Do not depend on a major infrastructure improvement just to do the prototype. This will delay development and will likely reduce the reliability of the prototype.

The first prototype is ready when you complete testing. Do not wait for users to test it. They will lose faith in the system if they see that it does not work. Test the parts that you know work and find areas where you know that failure will occur due to being incomplete. You will need to have users who can represent customers or suppliers.

Here are some steps to include before you announce the first version of the prototype:

- Complete the menus and navigation between menus and screens. This will provide an overall impression. Minimize the use of menus and use navigation buttons and links.
- Provide some limited user help to test the structure and format of this as well as usability.
- Develop several screens for entry and updating of information. This will allow users to carry out a transaction.
- To the extent that it is possible, implement the status codes and some of the workflow tracking capabilities.

Step 2: Review the First Prototype

Explain to department managers and staff their role in the development of the prototype—the use and review of the prototype. They will simulate part of the business activity with the prototype. They will also complete a review sheet for each prototype version, indicating issues and improvements. You want them to act like customers or suppliers. A side benefit is that they will feel closer to this audience through testing.

Here is a sample sequence of steps for the review:

- Explain the roles and steps to be followed
- Give a demonstration of the system, showing clearly where it ends
- Have the staff use the prototype for several hours as customers or suppliers
- Have the staff work with a live web site that is similar to yours

- Distribute forms on which staff members can provide criticism, suggestions, and recommendations as well as a comparison with the other web sites

On the suggestion form, include a line for the date, space to comment on ways that the prototype is better or worse than the current system and on differences between the systems, and a place for suggestions and comments. After each person works with the system, have him or her complete the form immediately. Leave the system for people to use after this test period, along with additional forms.

The first review provides an initial reality check. Are you on the right track? This is the first time that these people have seen something tangible. You may receive very definite opinions. This review typically sets the pattern for future work.

The first prototype should take about one month or less to develop. A review should take two to three hours.

Step 3: Work between Prototype Versions

Prototyping is a continuous process. At any time you have a list of potential changes. It is useful to categorize the list and to track it. By seeing the changing mixture of work you can detect a trend toward stability. Categories might include the following:

- Adding features or capabilities
- Modifying navigation
- Adding more user help and instructions
- Adding nonindex data elements and screens
- Fixing what was previously done
- Modifying the basic structure of the system

There is overlap in these categories. One change may fall into several categories. The last two categories are the most important, because correction or implementation may undo much of the work you have done.

What are possible actions for each item on the list?

- Implement the item in the next prototype. If you wait, you risk undoing even more of what you did.
- Set the item aside for later consideration. The item may be vague or there may be indecision about implementation.
- Defer the item if it interfaces to existing systems until the production system has been developed. There is no point in spending time on these items unless the prototype will be the basis for production.
- Renegotiate with managers and staff to drop, change, or replace the item. This is possible in the early stages of prototyping before people become too set in their ways.

- Review each item and act upon it as it arises. Leaving important items unresolved will impact the prototype. Pay special attention to workflow management, database indexes, and specific functions to ensure early definition.
- Group the list of items and changes by versions of the prototype. With each version of the prototype you will find more items to fix. Typically, you should include all changes in one area of the system for release in the same version.

How will you know if you are making progress? If the percentage of items to build in the prototype is diminishing, this is a good sign. The same rule applies to the number of errors that have to be fixed. No critical items should remain on that list.

Is the Prototype Being Used and Tested?

A lack of reported problems and issues is not a sign that all is well. The absence of issues often reflects a lack of testing. Or people may be doing the same testing again and again, so that nothing new is learned. There is a third possibility—that the same people are doing the testing and they have run out of ideas and become stale.

How do you prevent this problem? Examine the testing yourself. You can also plot several charts. The data from these charts originate from the testing of the prototype itself. Most large software firms conduct similar testing. You hope that the rate of finds diminishes with each new prototype. However, if a new version opens up many options, the find rate will increase. In addition, if different people test, the rate may also increase.

To ensure that there is adequate testing and feedback, rotate staff when you test the prototypes. Try to replace all of the people over time. You might also bring people back after several intervening prototypes. If you replace only a few people, you risk contaminating the testing, because junior people will become biased by what the existing staff indicate. Direct the testing. That is, indicate what is available in the latest prototype and where the staff should test. Then follow up in these areas.

How often should you deliver a new version of the prototype? It depends on the situation, but once a week or once every two weeks is often recommended. Some factors to consider are the following:

- The larger the time gap between prototypes, the greater the chance that the staff will become disconnected from the business activity.
- The fewer the number of prototypes, the less feedback you will have.
- Issuing versions too frequently means that items remain open. Users can get very frustrated because they did not get a chance to test the last prototype before another shows up.
- Agree on version frequency in advance so that staff members can plan their testing time.

Step 4: Complete the Prototype

Decide on the final version for the prototype. Not all open items will be resolved until the production phase. However, no critical items or major errors should block access to testing, even though functions may still be missing. The major criterion is that the final prototype be ready for pilot project use. You must be able to test the activity sufficiently using the last version of the prototype. The prototype should contain enough major functions so that the scope of testing of the new activity will not be limited.

**Step 5: Conduct the First Stage of the Pilot Project—
In-depth Assessment**

Employ a two-stage approach for the pilot project. The first stage is an in-depth assessment of the new business activity in the system using a small group of people. The second part is a test of the activity and system across a range of people and situations.

The processing of each transaction or type of work is examined. You usually begin with the most common situation, and then consider known exceptions. After this, validate that you have completely examined the business activity. Obtain a sample of work and talk to other departments to find other exceptions. In addition to coverage, you are interested in the following:

- Comparing the current and new E-Business activities in terms of ease of use and performance
- Detecting ambiguities in the new activity (i.e., gaps in procedures and policies)
- Testing new policies to see their effect on the activity
- Evaluating controls so as to detect errors
- Assessing the effectiveness of the measurement approach

To get this started, assemble a small team. Ask the team members to prepare a list of situations that the business activity should be able to address. Divide the situations among the members. Each team member describes in detail how the new activity should address the detailed situation. The result of this exercise is sometimes referred to as a *model.*

Document and test each model in the pilot project. This is tedious work that requires detailed business knowledge. This is one of the few times that you need dedicated effort by knowledgeable staff. The rewards are substantial. Not only do you test the activity, but you use the models as a basis for user procedures, training materials, and training. Thus, you achieve economies of scale and consistency. During this time, track issues that arise, just as you did in your work on the prototype. You can implement changes in either the activity or the system, or in both.

Step 6: Conduct the Second Stage of the Pilot Project— Wide-Range Assessment

In this second stage, the business activity and system will be employed by a wide range of people with different requirements. These testers will have a knowledge of the activity and background of the project. They represent a cross-section of the company. Not only will these people help refine the business activity transactions, but their work on the pilot project will get them interested in what you are doing. The approach will gain their support as they improve it.

Many failures in projects have occurred because the second stage was neglected. An automated teller machine model failed because it was tested only by those who had designed it. They never realized that it was too complex.

The goals of the second stage are as follows:

- Uncover the learning curve for the business activity
- Determine how easy it is to do work
- Solicit suggestions for improvement
- Obtain commitment and support for the new activity

Policy issues come to the forefront during the pilot project. Policies determine what people do and how they will do their work. The pilot project can create momentum for change and for testing policies. Through the new activity and system you find a better way to do something. At the start of the prototype, very few policy changes are known in detail. The prototype sheds little light on policy issues because the focus is on the system. Once you reach the stage of the pilot project, each model is mapped against current policies. You have to determine if the current policy can stand as is, should be changed (formal to informal, or vice versa), should be replaced, or should be dropped. If the policies are not changed in the pilot project, it will be difficult to change them later.

Once an issue is discovered, carefully define a new policy and evaluate it in terms of the old policy. Define precise terminology for the policy.

Here are some of the questions to be answered for a new policy:

- What is the impact of the policy on control and audit?
- How easy is it to circumvent the policy?
- How would someone know if the policy were violated?
- What should the scope of the new policy be?
- How should the company transition from the old policy to the new policy?

You can then prepare a written report of the following:

- New policy
- Corresponding old policies
- Differences
- Benefits and impacts

- Effect of the new policy on the new activity
- Effect on the new activity if the old policy is retained
- Transition approach
- Impact on other activities

Step 7: End the Pilot Project

The pilot project is finished after you have validated the workflow, transactions, and system. You may elect to continue it while doing conversion, infrastructure work, integration, and other activities discussed in the next chapter. Ending the pilot project is also a state of mind. Do you wish to freeze the workflow and transactions? Is it truly ready for production work? Have you demonstrated it to the right people?

Step 8: Document the Pilot Project and Prototype

You can now write up the survey results and demographics. Prepare a summary presentation to management highlighting the results of both the prototype and the pilot project to capture the essence of the experience.

One subject to cover is the team members. Do they work well together? Can they continue to work together in the next part of implementation? You can also document the differences between the old and new E-Business activities, as these are now in as sharp a focus as they will get. Revisit the comparison tables developed in preceding chapters and update them. Costs, benefits, and impacts can be revisited.

Finally, refine and write up the implementation plan for the remainder of the work, incorporating what you have learned about training requirements by job type and level.

Step 9: Build the New Infrastructure

While the prototype and pilot project efforts were progressing, work was being done on the infrastructure. Remember that changes here require the most time and money. Let's divide the discussion into infrastructure based on technology and infrastructure not based on technology. A basic guideline is that you do not want the infrastructure effort to get too far out in front of the pilot project because the pilot project experience may modify details of the infrastructure. On the other hand, you do not want the infrastructure to lag. If it is delayed, the momentum you established in the pilot project may be destroyed.

Technology Infrastructure

Technology infrastructure includes telephones, cabling, communications and networks, hardware, software, and staffing. Some specific concerns here are that much of the activity is sequential, decisions on technology are made too early, and changes are not synchronizing with the infrastructure. Some specific suggestions are as follows:

- Communications and networks are long lead items. Start them as soon as possible. They are often the least impacted by findings from the pilot project.
- Establish communications to handle the maximum you expect over the next three to five years. It is cheap to install additional cabling now, but much more expensive in labor if people have to return later. Labor is often the dominant cost.
- Delay hardware purchases as long as possible. Price performance increases and obsolescence mandate this. There are fewer economies of scale and discounts in purchasing the hardware at the start.
- Purchase hardware based on the experience of the prototype and part of the pilot project. Get more hardware power than you think you require because it may be expensive to upgrade and politically unacceptable to ask for more money later. Upgrades can be expensive in labor if you have to upgrade 100 computers.
- For certain main components, such as central hardware, hold back money so that you can buy more later when you have more experience with the requirements generated by the pilot project.
- Give a great deal of attention to integration and testing. The risk in technology infrastructure is typically not in the components but in integration.

Multiple Business Activities

Recall that you are implementing new approaches and improvements for an activity group. Much of what has been discussed applies as much to a group as to an individual activity. However, consider the complexity the group adds. Extend the prototype effort across the activity group. This will work if all of the activities in the group use the same system. The infrastructure work to support the activities in the group can be combined.

The difference in working with a group often occurs in the pilot project. It can be time-consuming and lead to delay if you have to undertake pilot projects sequentially. Here are some suggestions:

- Set up separate groups for the first stage of the pilot project, in which models are developed. Alternatively, you can use the same group of people for multiple activities.
- Develop specific models that cross the activities in the group.
- In the second stage of the pilot project, have some of the demonstrations oriented to staff members who work with one activity. However, also explain how the other activities in the group are impacted by the change.

Step 10: Integrate the New Business Activities

Integration is a general term that infers successful assembly of the parts. When you work on an E-Business implementation project, you cross many organizations, policies, infrastructures, etc. Implementing a new activity involves much more integration than the systems part alone.

Components of the New Activity and System

Integration attempts to mesh each step of the activity with the relevant part of the system. Integration also tests the interface between the staff members and the system. During integration, you reapply the scenarios of the previous chapter and generate user and training procedures.

The New System and Other Systems

Integration can be complex due to some of the following factors:

- The new system and the interfacing systems do not have matching data elements
- The different systems process information at different times so that there are compatibility issues
- One or more existing systems may be quite old and the data they employ may have questionable validity

Each of these factors may require additional programming for interfaces. An individual interface between systems can be one of several types. One is a batch interface in which one system produces a file and passes it to a second system. E-Business implementation often involves batch processes of orders for a short time period. This is the simplest to test. An on-line interface is the second approach. Dynamically, one system may pass a transaction to another. The third interface might be a manual one. Here, one person takes information from one system and enters it into another.

Factors considered in building and testing an interface are as follows:

- Verification of proper timing of the interface
- Exact data elements to be passed including their format and order
- Method employed to carry out the interface
- Confirmation that the data were received
- Method for verifying and editing data in the interface
- Approach for correcting any errors that are detected
- Recovery if the interface fails or is not present

The New E-Business Activities and Other Ones

The process interface includes the following:

- How information is tracked between the two activities
- Method for balancing and checking the interface between two systems
- Method for handling errors generated by the two activities
- Approach for obtaining combined information from the two activities

Here are some additional concerns:

- Consistency of procedures by the staff members who perform the work that tie to E-Business. If procedures differ, the staff may become confused, and processing will take longer. This will also create more errors. In some cases, this inconsistency has increased the time required by both activities.
- Policies that impact multiple activities. Conflicting policies are often a major problem. Different activities may be governed by different laws.
- Infrastructure shared by multiple activities. The concern is that performance of one activity will not interfere with that of the other one.

Activities and Policies

The pilot project covered policy impacts. The final new business activity and system must fit within the policies. In testing you will review the policy and then turn to the activity to see how smoothly and completely it follows the policy. If you want to circumvent the policy, will the activity allow you to do this?

Business Activities and the Organization

Analysis of how the activities and organization integrate leads naturally to thoughts on reorganization. You have already integrated a staff member with the new activity and system. At higher levels, you face issues such as the following:

- How will the organization supervise and manage the business activity?
- How will the organization deal with problems and exceptions that arise?
- How will the organization address policy issues and ensure policy enforcement?

Begin by defining the job requirements for specific functions of the activity and system. This will tend to redefine the jobs of people who do the work. Next, move up to the role and activities of the supervisor. Continue moving upward and toward generality.

Business Activities and Management

The integration between the business activity and general management moves you from the tactical handling of work and transactions addressed above to more general issues:

- How will management be able to control the entire activity?
- How will recurring problems be tracked and identified to management?
- What is the impact of management's decisions on organization and activity?
- What are routine management reports?
- What exception management reporting will be performed?
- What information does the process provide for errors, productivity, performance, cost, and staffing?

Not surprisingly, with all of your other work, this may be overlooked. People often just take it for granted that definition will follow. Here are some guidelines:

- Develop a strawman reporting method for the activity based on your knowledge of the business activity and system. A strawman is a model or sample of the reporting method. Put yourself in the position of a manager. If you ask people what they want, you may be disappointed. They do not know the process as you do and they may ask for much more than is possible.
- With the strawman defined, you have set the scope and format for integration. Once you refine this with examples and discussion, you can document it as a group of procedures to be employed with the activity.

Action 11: Convert the Information and Load Images and Information for E-Business

Often, when you implement a new activity and system, you must convert the information in the old ones. Data typically exist across the automated system, manual files, and even local files on microcomputers. Data setup and conversion can be the reason that the E-Business project flounders. If the data are not converted completely and properly, the entire new activity is jeopardized. Consider what might be wrong with the existing information:

- Information is faulty because there has been no update or validation.
- Images are of poor quality.
- Data in the manual and automated parts of the old activity are inconsistent.
- Information may originate in other activities and systems. These may have changed, contaminating the data.
- Data in the system are not in the proper format.
- Missing data must be found, captured, and entered.
- History data are in a different format than data of the active master file.

Even with modern technology, data conversion may not be a simple task. It may require extensive manual labor. After all, what would a computer program match against? In data setup, quality and completeness of information are crucial. Error-free transactions require accurate information at each step. Otherwise, additional manual effort may be required to find and enter the correct data, in which case productivity suffers. Poor data at the start may even deteriorate.

Begin with the new business activity. Determine the requirements for data setup related to sources of information, quality, and processing characteristics. Now review the results of Action 5 when you considered the old activity in detail. What problems surfaced related to data quality? Next, decide how to get the information into shape for the new activity.

Here are some factors to consider in your data setup decisions:

- Timing. If you transfer or reconstruct data too early, you will have to accommodate changes to the data prior to the new activity being implemented. This may require an entirely new, temporary updating activity. If you are too late, the conversion will delay implementation.
- The quality of the information. Test and sample the information to determine quality.

Step 12: Integrate the Parts of the Business Activities

Integration means putting together separate items, and then testing these items together. If A has 3 possible tests and B has 4 possible tests, then the number of possible combinations is 12 (3 × 4). Both normal and exception conditions must be tested in combination. You can never completely test everything, so some errors are latent and surface later.

Which items should you integrate?

- Loosely connected. An example is the employee and the computer system.
- Physically connected. The two activities are performed by the same people even though they are part of different activities.
- Tightly connected. The items are totally dependent on each other.
- Functionally connected. The items perform related functions, but are not totally dependent on each other.

The degree of integration and the type of connection affect the extent of testing required. The more integrated the activity is, the greater the likelihood that the activity is effective. Reduced effectiveness relates to the activities being loosely connected.

In planning the integration testing, take advantage of loose connections between parts of the activity and systems to reduce the amount of testing required. Also, systems integration take place behind the scenes while activity integration tends to be more visible. This generates a wider variety of tests.

Consider different types of testing. Begin by finding the major parts of the interface that have risk in terms of their impact on the business activity. Then determine the type of testing required. Here are several types:

- System testing. In this type of testing, only the automated part of the activity is tested. The overall process is not tested.
- Process, nonsystem testing. In this case, the activity is tested. At the points at which automation occurs, insert a black box to avoid system testing. This is suitable for exceptions.
- Transaction testing. Here a transaction is taken through the entire workflow (including both system and manual operations).
- Performance testing. In this type of testing, you see how the business activity and system behave under stress loading. You can determine throughput (volume of work over time) and response time to do the work.
- Acceptance testing. This is testing by the department staff and management to see if they will approve the entire activity acting as the customer or supplier.

For E-Business you have to carry out testing of the content of the web pages and their appearance. For example, if you are selling an item, you have to check the description, price, shipping and handling information, part number, color options, additional options (such as monogramming), and weight. This is essential but tedious work.

Decisions related to the test approach include the following:

- How will you obtain the data? Will you use standard production data from the current activity? Will you collect data separately?
- How will you test? Go back to the scenarios and build test scripts. A *test script* is a set of procedures for doing a test. The script identifies the anticipated result. People are reluctant to spend time creating scripts, but they should at least use their scenarios from their pilot project efforts. Scripts are beneficial because they provide formal structure for testing.

Next, carry out the following steps:

- Set up the activity and system. Purge old test data from the activity, set up a computer test environment and process test environment, and assemble staff for testing.
- Carry out testing and monitor the testing. Often, you can gain useful information from observation of the staff carrying out the work. The people doing the work may miss details because they are concentrating on the workflow. Also, conduct short interviews during the testing. Do not hesitate to stop the testing and restart it.
- Analyze test results. Tabulate the portion of the results that is quantitative. Look at the subjective opinions and reactions. You will find new informa-

tion that may contradict lessons learned in the pilot project. Determine not only what to do but also why this happened. Was there a problem in the pilot project? Was it carried out for too narrow a scope? Generate changes in the procedures as well as error reports for correction to the system.

- Network. The network design and communications hardware may not be able to handle the load. Like PCs, communication hubs, routers, and gateways come in a variety of models with different performance levels. In addition, the network design typically divides the body of users into segments. Your new system may overload a segment.
- Databases and hardware. The bottleneck could be the server hardware, operating system, and database workload.
- Computer-based monitoring software. This software can indicate problem areas and where the system resources are consumed. In terms of countermeasures, you can upgrade or modify networks relatively quickly. It takes more effort to modify and optimize the database requests. Hardware upgrades can be expensive, but they are generally routine. If you have to redesign the database, the project may be in trouble. After you find all of the issues (or as many as you can), fix them in parallel.

Step 13: Choose and Implement Backup and Disaster Recovery Methods

Linked to, but separate from, history data are the issues of recovery and restart. If you need to resort to backup, how will you recover the activity and system? How will the data that were generated or obtained while the backup was in use be loaded or transferred into the activity? Much has been written about system backup and disaster recovery, but what about the activity? *Restart* means starting the process and system up again so that no transactions are lost or processed in error.

How can a business activity fail? Some examples are as follows:

- The web site becomes very busy during peak times such as holidays so that the system crashes or at least response time is very bad.
- The system fails, causing the activity to fail. If the software fails, the site is gone.
- The web site is not easily usable, causing customers to go elsewhere and suppliers to use manual channels.
- People with extensive activity experience may leave, causing the activity to deteriorate rapidly and fail. This may even occur in E-Business since you are still dependent on internal staff for critical knowledge.

The first step in addressing backup is to isolate the few critical activities— those that you cannot allow to fail. Having identified these, add in activities on which the critical activities depend. For other processes, data will be collected, but the information will not be processed.

Proceeding now with the critical activities, determine the risks to the activities, not just to the systems. There are wonderful disaster recovery plans for major systems; however, they often do not include getting critical staff to work to use the recovered system.

Some considerations in developing your backup plan are as follow:

- What is the minimum level of a business activity in terms of operation? Can only one or two functions be supported? For example, for an insurance company, billing and claims would be considered more important than new application processing.
- How long will it take to establish the minimum acceptable level of the activity? What happens with the data, customers, and so on during this period?
- After you define the minimum activity, what is the recovery sequence to restore the activity?

The simpler the activity, the easier it is to bring in backup staff to do the work. Also, the more modular the work, the greater the flexibility in placing work for processing.

Step 14: Construct On-line Activity Procedures

Your new E-Business system will require on-line help and procedures. Some of this was developed during the prototype and pilot—it will be finished here. Additionally, internal staff may be involved in the activities so that they need procedures. In the past, organizations documented activities with procedures. Procedures for the computer system were then separately documented. In E-Business you want to have a unified set of procedures. The procedures should be prepared by department staff members who will work with the activity and customers or suppliers. By participating, these staff members will gain a sense of ownership. The procedures will be in their language—the same as that of the customers and suppliers. Also, their participation increases the chances of the procedures being used.

Step 15: Train the Staff

Even in E-Business you have to train staff to support customers and suppliers. Develop training materials when you develop the activity procedures. The materials should draw directly from the scenarios of the pilot project as well as from test scripts. Include an overview of the activity, as well as detailed training exercises.

The following guidelines have proven useful:

- Always provide an overview of the activity and system to participants at the start of training. Include parts of the activity that are not within their job descriptions.
- Link the E-Business activities to the policies.
- Describe what management expects from the new activity so that people will know what they are to do in terms of performance.
- Include hands-on work with the new E-Business activities.
- Ensure that the cut-over to the new activity occurs soon after the training is completed. Any gaps will cause problems in retention of the activity knowledge.
- Use department staff extensively in the training.
- Consider using the train-the-trainer approach so that there is wider involvement.
- Share war stories and experiences that support performing the activity properly.

Step 16: Cut-Over to the New Activities

In the past, a variety of approaches have been proposed for cut-over to the new activities. Three of these are as follows:

- Parallel. Operate the new system in parallel to the old system. At an appropriate point in time, terminate the old system. In E-Business you would continue to work with suppliers using the old process.
- Pilot project. Install the new system with one group of customers or suppliers in production (not in the pilot project mode discussed in the previous chapter). Then expand it.
- Cut-off. Simply stop the old system. The new system takes over.

You really should get people's feet wet through an initial experience. Therefore, we recommend the pilot or parallel approach since these are the most conservative for E-Business.

ORGANIZATION AND MARKETING IMPLEMENTATION

Define and Implement the New E-Business Marketing Structure

The new marketing structure will typically have the following elements to support E-Business implementation:

- Group responsible for analyzing sales and web use data. This group will have to be provided with suitable statistical tools.

- A group or individual who will be responsible for checking out the competition and what other companies are doing on the web.
- Promotion and discount coordination between the web and traditional channels.
- Strategy review and updating where marketing constantly reviews its approach to E-Business.
- A group that analyzes customer demographics and purchasing on the web.

Implement Changes in Other Departments

E-Business for retailing has impacts on accounting. Accounting in many firms has not had to deal with credit card transactions. The accounting department will probably have to establish a group that deals with the following:

- Credit card sales reconciliation
- Credit card disputes
- Accounting relating to returned items and refunds

Finance will be involved in setting up management reporting on the web sales and activity. In a retail environment, shipping and fulfillment may need to change to adapt to E-Business, where shipping must be fast. IT is also touched since there must be a rapid response unit and capability within IT for E-Business problems.

Implement Changes in Other Business Activities

In IT there must be a new method for working with marketing on promotions and discounts. This will give IT a heads-up on what new requirements are coming. The warehousing and inventory policies may have to change to support higher inventory levels to prevent back orders from becoming critical. Accounting will have to establish new activities that support the functions listed above.

Changes to the Organization

You may want to wait until E-Business is firmly established before proceeding with this. However, it is not too early to do some planning in the E-Business implementation. The impact of E-Business activities should become clear through the pilot testing and evaluation of the activities. You can be helped through the comparison tables that you constructed earlier.

To begin planning, define the organization structure and assignment of staff into the various roles after relating the positions to the activity. You can continue bottom-up and associate names with positions in an organizational chart that you created with the activity fit to the roles. This method detracts from an opportunity for organizational improvement. A second approach is to now move top-down. Define the scope of departments based on the activities as they have been redefined. As you establish the overall organization, you can fit together what you have done from top-down and bottom-up.

Assignment of staff to roles should not be left solely to the managers. The implementation team has acquired experience and knowledge of the staff. This can be useful input. Each level of the organization near the activity can be populated by the appropriate managers and staff.

As the cut-over approaches, a lack of fit between the current organization and the new E-Business activities may become evident. Organizational change is needed. Here are some options on the timing of such change:

- Prior to cut-over. This period will be disruptive and may prevent successful implementation.
- As part of the cut-over. At this point, it will be enough to get the new activity going.
- Shortly after the cut-over. This is probably the least disruptive period.
- Long after the cut-over. At this point, it will be too late, as the old organization has seized on the new activity.
- Independent of cut-over. This period appears feasible, but it negatively affects integrated change, the overall purpose of E-Business implementation.

PRESENTATIONS TO MANAGEMENT

You cannot avoid this since management will be quite nervous. When you show the pilot project to managers, have the business department staff do the entire demonstration. Introduce it and then move to the sidelines. The staff members' passion and conviction will influence management. At the end of the demonstration, return to the stage and go over the plans for the next step and review the costs and benefits.

When discussing the costs and benefits, be direct. If it will cost more or have fewer benefits with what you have learned, admit this. Money and time have been invested in the prototype and pilot project, as well as in the infrastructure. This fact, combined with the enthusiasm of staff members, will likely influence management to continue. The cost from this point on is limited.

E-BUSINESS EXAMPLES

RICKER CATALOGS

Rickers followed the development and implementation approach but failed to do adequate testing. There was some system testing, but a lack of user acceptance testing. This resulted in the system crashing often in the beginning.

The system was brought down, fixed, and then extensively tested. After that testing was always a part of change.

MARATHON MANUFACTURING

Marathon followed the steps in this action and allocated sufficient people for testing that over 400 errors and changes were made during testing. Remember that this firm is an exception because it faced less time and money pressure.

ABACUS ENERGY

Initially, Abacus felt that they did not have to provide extensive on-line help. After all, the number of transactions was small. Errors began to occur in production so that on-line help was added. Next, they did not believe that they needed to review the content of the material put on the web for request for proposals. They found that the paper format for this did not work and went to an entirely new design that was more modular with links.

CRAWFORD BANK

Crawford provided adequate testing resources and followed a stringent quality assurance program. They also reviewed the web sites of many other banks and institutions. This was useful in both the design of the prototype and the testing.

E-BUSINESS LESSONS LEARNED

- **Instruct staff members to focus on substance rather than appearance during the prototype and pilot project testing.**
  Make sure that people focus on how a specific transaction is addressed in the system.
- **Look for new opportunities for E-Business to surface during the pilot project.**
  Do not ignore these. At this point, write them down. Wait until later to work on them. Avoid promises until you have thought through the opportunities.
- **Alert management as soon as you discover the need to develop a different E-Business approach.**
  The pilot project may lead you to the realization that the new activity is not going to meet the issues and underlying problems without organization change. This should be explained to management early.

- **Business activity implementation must accommodate or address issues of style.**
  In E-Business you must be culturally sensitive if you are marketing in different countries. The different web sites must have individual content.
- **Use testing for political advantage.**
  "Test results are viewed as academic information to be used for technical work." If you believe this, you are mistaken. If department staff have been involved in testing, and results are favorable, ask these staff members for testimonials.
- **Consider using parallel implementation or multiple implementation.**
  If the implementation of a business activity is successful, consider implementing multiple additional activities at the same time. Thus, you will reduce the number of separate changes and provide economies of scale of implementation for the team. The dangers are diffusion of resources among multiple activities and confusion among department staff. In many cases, the balance is on the side of multiple activities.
- **Delegate responsibilities to departments as soon as possible.**
  During E-Business implementation, the team may be tempted to hold control over the activities. Instead, hand over control as soon as possible. The implementation team should move into the background and monitor the work.

WHAT TO DO NEXT

1. Identify the people who will be involved in the prototype and their roles.

   Individual Role

   _____   _____

   _____   _____

   _____   _____

2. Identify the people who will be involved in the pilot project and their roles.

   Individual Role

   _____   _____

   _____   _____

   _____   _____

3. Write down the requirements of the prototype in the left column. In the right column note the specific results or conditions under which the requirement is met.

Requirement Condition

_____   _____

_____   _____

_____   _____

4. Track each successive prototype in terms of the factors listed below. This table can be employed to determine trends toward stability. The time lag in this section refers to the elapsed time between the issuance of the previous prototype and that of this prototype. The major change is the major improvement that was included in the prototype. In the final column, indicate what remains to be done. If the last two columns do not change, there is a problem and the developers are likely concentrating on minor enhancements.

Prototype Number	Time Lag	Major Change	Left to Do

5. How does the current set of employees fit with the old tasks? Write down the names of the people in the rows. Put the tasks of the old activity in the columns. In the table, use the rating system of 1 to 5 (1 means the person does not perform the task at all; 5 indicates that the person is almost totally responsible for the task).

Job Titles

New Activity Tasks			

6. Determine how the employees would fit with the new activity. Enter their names as rows and the tasks of the new activity as columns. Use the rating system of 1 to 5 (1 means the person is not suited at all; 5 indicates that he or she is

a very good fit) to indicate the degree to which you or others feel they are suited to the task.

Job Titles

Old Activity Tasks			

7. You can now use your analysis from Action Items 1 to 4 and create a new table of people versus job titles. Use the rating system of 1 to 5 (1 means there is no fit; 5 indicates an excellent fit). This table shows you possible candidates for specific jobs in the new activity.

Job Titles

People			

Chapter 15

Action 13: Follow Up after Your E-Business Implementation

INTRODUCTION

You are finally in E-Business production and the new E-Business activities work. Unfortunately, people are often now so tired that they do not even want to expend the effort to measure the results. However, if you miss this opportunity, it will be difficult to return to gather data later. You will have lost your rapport with the staff. You will probably not have enough time to gather data later because you will have moved onto other E-Business efforts. It will be harder to collect information because you will have to reestablish contacts. In addition, there will be a tendency to be complacent from a business view. However, competitors and startup firms are not ignoring this. Nor is the investment community. They will want to know how you are going to expand your E-Business.

ISSUES DURING E-BUSINESS OPERATION

Here are some issues that can arise after initial E-Business implementation:

- New problems surface in the organization. There may be marketing problems in dealing with E-Business. There could be problems between IT and marketing or other departments.
- Customer complaints may be quite numerous at the start. This may raise concerns about whether E-Business was a good idea.
- The benefits of the new activity are not measured, but the old and new are compared. People forget the old methods. The benefits of the new activity

can now be questioned. This is likely to occur in the initial stages of E-Business when traffic is low.

- People question the method itself. With no measurement of the business activity, people may question the workflow and procedures because the default visible measurement is cost (which has a negative connotation).
- Surrounding systems or activities change, forcing an unplanned change. This is why you must plan changes and enhancements to systems and activities on a department, rather than individual, basis. In E-Business there can be stress in shipping and backorders, for example.
- Staff members who were with the new activity when it was implemented may leave. They usually do not transfer their knowledge prior to departure, and the remaining knowledge is limited.
- New exception situations arise. No one addresses them formally, so staff members begin to develop procedures of their own, which work around the activity. The entire business activity is now jeopardized as more work moves outside of its framework.
- The organizational change on which the business activity depended did not occur. The activity now does not fit the organization, so the organization changes the activity to fit the organization. This is a real problem in E-Business where organization change lags due to the pressure to implement E-Business.
- Other activities, independent of this one, are changed by other teams. The other new activities are not compatible with this new one, so potential benefits are not realized.
- Competitive pressure and the business force change. No one improves the E-Business activities to respond to the new conditions. Management is frustrated and creates a parallel workflow for the new work. Lessons learned and success experiences are not transferred.
- Technology changes, but the staff are stuck with the old technology. Morale sinks as old technology continues to age. New e-commerce software or hardware emerges and the IT group is not pushed to evaluate it.
- A better way emerges. The new activity triggered creative thoughts, but no one formalized changes as part of the work. This results in work variations. People learn from E-Business and now want many more changes.

Not only should you respond to these situations, but you should also anticipate and monitor the regular business and E-Business activities and their surroundings for signs of pressure and change. In traditional information systems, the technical staff responds to user requests for changes to the system. When you move to E-Business, you learn that you must be more proactive.

E-BUSINESS MAINTENANCE VERSUS ENHANCEMENT

E-Business maintenance for an E-Business activity means that you take action to ensure that the E-Business activity continues to meet its goals and requirements. Maintenance refers to being able to accommodate situations with the current E-Business activity. Examples include taking steps to reduce response time or errors to acceptable levels, and continuing to provide the same information due to high customer or supplier volume. Maintenance involves updating product and service information on the web. Here the software is not changed, but the content is. If you cannot keep up with getting new products on the web, then immediately you must improve the new product setup activity for the web.

An E-Business activity may be forced to respond to new pressures and require-ments. In the case of Crawford, other banks and new banks entered the world of E-Business lending. Crawford had to push out new features and capabilities as well as product variations. Meeting new requirements or attaining new perfor-mance levels is referred to as *E-Business enhancement.* This may entail substantial changes. An enhancement may apply to an activity, organization, or infrastructure change. Examples of E-Business activity enhancements are the following:

- Implementing new types of products. An example is Amazon.com's move into new products beyond books and music.
- Implementing new features such as lessons learned and instructions. Mara-thon's effort to put in a wizard for machine shop job bidding is an example.
- Processing a new type of exception transaction. Crawford implemented new methods for handling exceptions.
- Obtaining additional customer and supplier statistics from the E-Business activities. Ricker found that they needed to have more customer information to determine what direction new products should take.

E-Business enhancements tend to be proactive. That is, a requirement surfaces and the E-Business activity is enhanced to meet the requirement according to the overall E-Business strategy. Maintenance, on the other hand, tends to be reactive. The mixture of maintenance and enhancement work then is an indicator of the state of the E-Business activity. It is a sign of health or decay of the business activity.

Some observations on this subject are as follows:

- A lack of maintenance and enhancement may not indicate success; it may indicate that the business activity is not being used.
- The greater the percentage of total effort going into E-Business enhance-ments, the more the activity is being used. This is because people expect more from the activity and have defined additional functions.
- Excessive E-Business maintenance may mean that people are attempting

to adapt the business activity to new work without change. A good effort is being made without success.

- E-Business enhancements are easier to measure because they require resources and, hence, management approval. Maintenance can often only be detected through observation since support is provided by the business and IT groups.

Here are the objectives in this action:

- Ensure that benefits of the new E-Business activities are achieved.
- Defend the integrity of the E-Business activities from attacks and degradation. If the IT group just patches on changes to the software, future change becomes more difficult.
- Use the information from the E-Business activity as a direction for future change.
- Explore and capitalize on investments to make further improvements.
- Proactively enhance and support the integrity and performance of the E-Business activities, their infrastructure, and their organization.
- Use success in the implementation to move on to new E-Business activities that will be implemented.

MILESTONES

There are several end products in this stage. First is an assessment of the benefits and costs after implementation to be presented to management. Second is identification of other possible areas of expansion and improvement. Third is the application of lessons learned from the project to other work. The overall end result is a dynamic E-Business.

METHODS FOR E-BUSINESS

There are two aspects to the ongoing work after implementation. One is to address the E-Business itself. The other is the support of the E-Business activities.

E-BUSINESS MANAGEMENT

How Are the New E-Business and Regular Activities Performing?

One answer to this question lies in the analysis of the performance of the technology and systems. Here you are looking at responsiveness of the systems to peaks of web use. Another answer centers on the business side. Here are some more detailed questions to address:

- What is the level of customer service being provided?
- What are the nature and number of complaints being received?
- How easy is it for customers or suppliers to navigate the web site?
- What is the volume of sales or work being performed via E-Business versus traditional business?
- What is the cost of transactions performed through the web?
- What are competitors doing with their web sites? What else can you do to improve your competitive position on the web?
- What is the performance of the systems in handling peak volumes of customer or supplier activity?

How Are the Information and Experience Being Employed?

E-Business generates many lessons learned through experience. Marketing learns what is effective and how to be more dynamic. Other business departments learn how to cope with E-Business and standard business at the same time. You need to have some method to capture and use this information and experience.

Is the Organization in Tune with the New E-Business Activities?

Is management changing their way of thinking about their business? Or, are they viewing E-Business as just another channel. If so, they may not be responsive to change. E-Business activities should force management and the organization to think more creatively (the trite expression is to think outside of the box).

E-BUSINESS ACTIVITY EXPANSION

The steps here are in no particular order. Some of them are optional, depending on the state of the E-Business implementation.

Step 1: Appoint an E-Business Activity Coordinator and Monitor

Critical E-Business activities are so important to the business that organizations should create the role of the activity coordinator and monitor. You can begin by identifying a staff member for a group of related E-Business activities. This position would be part-time so that you might also appoint a primary and secondary coordinator. The duties of the coordinator include the following:

- Collect information on the business activities and issue an activity report card.
- Survey the industry as well as analyze internal data to discover new E-Business opportunities.

- Review infrastructure and organization issues related to the business activities. (This includes requests for changes as well as analysis findings.)
- Recommend and design specific changes and improvements to management as well as alerting management to problems.
- Manage the implementation of change.
- Coordinate the training of new staff in the activities and collect the experience and knowledge of departing staff.
- Coordinate the change to computer systems supporting the E-Business activities.

All four of the example firms appointed several E-Business coordinators. This is not a job for the department supervisor because business activities may cross multiple departments and the supervisor is in only one department. The supervisor may have vested interests in the organization and may not be loyal to the specific activity. Also, the supervisor may have too many other duties.

There are two levels of coordination in business departments. The first level is the tactical day-to-day operational coordination of departmental work. This requires detailed business activity and systems knowledge. The second entails a wider area across organization and policies. At the second level, the coordinator may have less hands-on knowledge but be more politically savvy.

Choose a middle-level or junior staff person with experience in the systems and business activities for the role of E-Business coordinator and monitor. Such a person is more interested in change and less tied to the old ways. You may get double duty out of this person by also having him or her act as a system change coordinator. A more senior person could be the alternate, second coordinator. Measure the activity coordinators by the performance of the business activities and the quality of their measurement and reporting. Encourage E-Business coordinators to get together to share ideas.

The existence of an E-Business coordinator does not reduce departmental accountability. Business activity measurement includes the extent to which each department is performing its part of the activity and its quality.

Step 2: Fill Out an Activity Report Card

An activity report card (an activity evaluation) is completed by collecting all of the measurement information. It is important that there be multiple report cards, one for each activity in the activity group, so as to provide a comparative basis.

Here are some items to include in the activity report card:

- Overall grade/evaluation
  — A: Acceptable; no improvement required
  — B: Activity works, but improvements (enhancements) are possible
  — C: Activity works, but benefits are not being achieved

- — D: Activity does not work all of the time; there is a high error rate
- — F: Activity fails and is not acceptable
- Customer acceptance and comments on the service and the web site
- Specific grade components
  - — Organization
  - — Staff
  - — Systems
  - — Technology
  - — Infrastructure
  - — Error rates
  - — Resources consumed by business activity
  - — Throughput
  - — Response time
  - — Subjective opinions
  - — Open business activity items
- Objectives for the business activity
  - — Short-term
  - — Long-term
- Assessment of the E-Business activity vis-à-vis the competition and industry

To implement the report card approach, first identify the E-Business coordinators. Next, coordinate the assessment of multiple activities for E-Business and regular business. This will give you a sample set of report cards. This is referred to as the *activity report card baseline*.

Step 3: Generate an E-Business Activity Semiannual Report

Organizations produce annual business reports, organization charts, and other similar documents. Surprisingly, they do not report on business activities. In many organizations a semiannual activity report is encouraged due to the dynamic nature of E-Business. This report basically contains graphs and findings for critical activities. It also relates the work of the next six months in terms of projects to the list of critical activities. In one retailer, among the annual activity report, the technology assessment, the competitive assessment, the information systems plan, and business strategies, the semiannual activity report received the most attention.

The semiannual activity report provides information to management and staff on the state of their E-Business activities. It is neutral and so very valuable. If the information is faulty, you will hear about it quickly and you can learn from this feedback. You will also see that after the report is issued, more of the systems, organization, and infrastructure documents will tie directly to critical E-Business activities.

Step 4: Evaluate E-Business Costs and Benefits

In E-Business people say that you should not expect to make money immediately. While this is true with startup firms, an existing business should push for tangible benefits and measurement of costs and benefits. The bottom line is that E-Business has to be justified economically.

The typical question posed in a postimplementation review is about costs and benefits. In the past, you probably defined a long list of benefits ranging from economic to organizational impacts. The situation can change when you reach an operational state. People are less concerned with impacts, because the effort represents sunk costs; in any case, you will have to live with the business activity. In addition, the relative weighting of the benefits and what you considered important have changed.

The categories of benefits resemble the three standard levels of management and are as follows:

- Operational. The focus here is on the functioning of the E-Business activities.
- Managerial. The attention here is on economic benefit.
- Strategic. Consider benefits to the organization, policies, and general infrastructure.

Operations

The first hurdle is operations. If the new business activity did not successfully create a new business or augment or replace the old activity, there is no point in measuring higher-level benefits. Questions to answer are as follows:

- Does the new activity handle the workload?
- Is the performance satisfactory in terms of volume, response time, and error rate?
- Is the new activity stable? Or is there fluctuation in performance?
- What are the managers' and employees' opinions on the activity?
- Does the new activity address all exceptions?

Management

Now you move to the more traditional cost–benefit analysis. Consider first only tangible benefits. Intangible benefits should be reflected in operations and strategy. Here is a partial list of the information you want to determine:

- Revenue-related
  — How much sales volume has come through the web?
  — How many business transactions were shifted to the web?

- Cost-related
  — Operating costs (direct staff and organization costs for the new and old activities)
  — Support costs (technology, systems, and infrastructure costs for the new and old activities)
  — Development costs for the new activities
- Performance-related
  — Error rates and amount of redoing work for the new and old activities
  — Volume handled for the new and old activities
  — Staffing levels for operation for the new and old activities
  — Staffing levels for support for the new and old activities

After you have gathered the necessary information, you can then perform the standard cost–benefit analysis.

Strategy

E-Business implementation should do more than yield a good report on costs and benefits. Go beyond the standard impact analysis. Return to the comparison tables you created for the new E-Business activities. Create updated tables and put them side-by-side. Address the following questions:

- Do the new E-Business activities and changes support the business vision?
- Does the E-Business implementation change the organization and infrastructure as desired?
- Do the E-Business activities support flexibility in the business?
- What are the impacts of the new E-Business activities internally on morale?
- How do the new E-business activities help management direct work better?
- Do the new E-Business activities have an impact on the external community (e.g., suppliers, customers, investors, and regulators)?

The answers to these questions are subjective. You can survey managers for their opinions. Ask yourself what has changed. You can also consider the vision, management, external impact, and other factors and see if you can broaden their impact because of the new E-Business activities.

Intangible Benefits

You have looked at tangible benefits. However, you will find a number of intangible benefits that people agree are present and consider to be benefits. The problem is how to use these intangible benefits without affecting or detracting from the quantitative benefits. Here are some examples of intangible benefits with suggestions on how to quantify them:

- **The new business activity is easier to learn and use for both employees and customers/suppliers.**
  There may be fewer errors. Training time and costs should be less. Productivity and morale may increase.
- **Information is easier to find and is accessible.**
  More information is typically on-line in E-Business. Transactions are more visible. Look at how the time for retrieval has lessened. The number of complaints from customers when you could not find their information may have dropped.
- **More history information is available than before with the old activity.**
  In E-Business you have not only the detailed transaction, but also all of the navigation that a customer or supplier performs. Look at how much time was spent in the old activity on retrieving and reconstructing history compared with the new activity.

Reviewing the Costs and Benefits

Make sure that you are not the only person who can support the costs and benefits. Have a direct source for the information—even opinions. To get opinions and find people to give testimonials, ask people what they would do if they returned to the old business activity. Problems with the old activity (which the new activity lacks) will surface. Look for customer and supplier comments as well.

Work Left Undone

In reviewing benefits, you will probably notice items that have not been completed or closed out. There still may be exceptions and shadow systems. In fact, new shadow systems may have had to be created to respond to specific situations such as discounts that could not be handled by the systems.

Step 5: Plan Your Next E-Business Actions

Where do you go next with E-Business? Do you expand products with the same capabilities, or do you expand the features and range of products on the web? In your plan and strategy, you formulated an approach and direction as to which business activities to address next. Implementation and politics could have entered the picture. It is time to revisit this issue and negotiate with management to go to the next activity group. A manager could have been favorably impressed with your efforts. The requirements of the business could have changed so that you must alter your focus and address activities you had originally planned to ignore.

Step 6: Detect and Remedy E-Business Activity Deterioration

As the E-Business activities age, negative changes may occur. Software increases in complexity and size as it ages. Infrastructure supporting an activity may not be maintained. Technology becomes obsolete and must be upgraded. The activity itself can degrade in many ways, including the following:

- The systems may not be responsive to new demands by marketing
- Infrastructure deterioration affects the activity
- Technology deterioration affects the activity
- Organization change affects the activity
- More of the work is performed outside of the activity
- Procedures within the activity are not followed
- Different people use different procedures

Business activity deterioration can be summarized as declining competitiveness. You are worse off because you raised management's expectations and wasted resources. You also destroyed a future opportunity because people will now be reluctant to revisit this.

Some of the recurring examples of business activity deterioration include the following:

- Competition heats up. The competition implements changes and new features that overcome what you have done.
- Manager replacement. A new manager changes the activity to reflect his or her style. The manager made his or her mark, but the changes were not thought through and the activity decayed rapidly. That was fine with the manager, who then moved on to another disaster.
- Loss of critical staff. Back in Action 5 it was suggested that people who know a business activity well can still support it. Unfortunately, their departure typically goes unnoticed.
- Systems and infrastructure deterioration. Sometimes the computer system or building decays to the point where the activity performance is affected. When a business activity begins to decay, the rate of deterioration quickens. Another case occurs when the software is not modernized.
- Inadequate response to change. Management uses a stopgap measure that ends up being in place for years.
- Organization change. The organization changes through downsizing and outsourcing. Qualified people are gone. The remaining staff may not know the activity. At one company, employees saw that layoffs would be coming, so they destroyed the procedures. Then they could not be terminated.
- Old business activity remnants. Individuals continue to use the old activity but pay lip service to the new. The result is that people are using a mixture of the two activities.

- Conscious effort to return to the old business activities. Management and staff in the department work to move the activities back to the format of the old ones. The activity is old; the system is new.
- Lack of orientation for new staff. Once a new business activity has been implemented, the departments may not train new staff in the new procedures and E-Business. Someone assumed that there would be no turnover, or that people would learn the activity through hands-on work. Thus, new employees are dumped into the activity. You really need to give E-Business training to the new staff.
- Power shift. Power positions can change quickly. A manager in another area may attempt to steal staff or take over the functions of E-Business and regular business. In some cases, an entire regular activity can be lost if you lose the critical people.
- Emergence of shadow systems. These may be workarounds to the new activity or be created to support new work that no one could put through the new business activity.

How to Detect Business Activity Deterioration

Here are some guidelines for detecting deterioration:

- Review customer or supplier complaints about the activities and web site.
- Review how the E-Business activities are supposed to function. This gives you a baseline of what is supposed to be done.
- Review any measurement information on the business activity. Data might be available on staffing, volume, response time, and error rates. You are looking for variances.
- Determine if there have been any staffing changes. Can the time of the change be related to the measurement data? Identify the new staff.
- Contact employees outside the business activity who use output from the business activity or feed it. Determine any issues or trends that they detect.
- Contact employees in the department and observe the business activities. Consider how exceptions are handled. Go to the key people and get their opinions. Ask new employees how the activity works and what they find wrong with it.

The purpose of this effort is to detect opportunities for improvement and problems. Use the suggestions from Action 5. There are three types of information: quantitative, observational, and perceptual. Rather than separating these, concentrate on using all of the information to address the following:

- What change has occurred? Why?
- How could the business activity be improved more?
- What is the sequence of actions that are appropriate for the future?

Alternatives for Action

Some alternatives for action are as follows:

- Do nothing. Let the business activities drift and continue to decay. Do not reject this option out of hand. Often, things have to get bad before anyone intervenes. This method has been used in some companies with management's approval. A trigger can be defined as to when a crisis level is reached; management then has justification to act without political risk.
- Remedy the situation by tweaking only the activities. This is the after-the-fact temporary approach. This approach is popular with managers who want to move on to another job and do not want to see a major project through to conclusion.
- Develop a plan to improve the E-Business activities again. This would include buying new technology and developing new systems and infrastructure. Do not expect much enthusiasm for this alternative. People have spent money on the business activity and do not want to continue to do so.
- Adjust the organization and policies. This is probably the most feasible approach if you want affordable change.

Doing nothing is not as bad as it seems. If you attempt to tweak and change the business activities too often, you threaten its stability. You also consume valuable political points.

Step 7: Intervene to Fix E-Business Activity Problems

Intervening in a business activity can have several goals. The major goal should be to implement E-Business activities of self-correction and self-policing. This means that you will not have to revisit the E-Business activities again and again. Another goal is the assertion of control over the E-Business activities. You can actually accomplish both goals. If you cut through middle management and work directly with the employees on the business activity, they become involved and dedicated. They align with you and your control rises. You can also pave the way for eliminating middle management.

Some of the risks of intervention, along with suggestions for risk minimization, are as follows:

- Demoralization of staff. The morale of the people involved in the activity drops. How you intervene and what their roles are will be important factors.
- Underlying issues. If you attempt to intervene, you must address basic organization and infrastructure issues. A fundamental flaw in the new business activities may require basic change. Unless there is a crisis, there is a lack of will to make the change.

- Raised expectations. People think that you will carry out a major, cheap, effective short-term fix. This appears to happen often, but seldom really does. Often, some new headline issue grabs people's attention.

Intervention Approach for E-Business

Consider carrying out the implementation in stages. First, you must understand what is going on currently. Collect a limited amount of information and determine the initial action desired from your alternatives. Because you may review multiple activities, you cannot afford to spend a great deal of time on every activity. Instead, allocate your time. Intervention has both a short-term and a long-term focus. The short-term focus is to fix symptoms of problems, while the long-term focus is on the problems themselves.

Assuming that you prepare an activity report card every six months, you might construct the following graphs:

- Revenue over time
- Cost, performance, and volume over time
- Ratios
- Cost per transaction in the business activity
- Average time per transaction in the business activity
- Error rate versus staff turnover
- Division of cost of work over time—staff, infrastructure, technology, etc.

How can you use these graphs? For one activity you can detect trends in deterioration and performance. You can use them to enhance interest and knowledge in the E-Business activities. Perhaps the most important use is comparison of multiple activities.

Suppose you identified 10 critical activities for the business. Using the measurement information, you can develop comparative graphs on business activities. You can use charts like this to do the following:

- Compare activities that are not easily compared (accounting versus inventory)
- Identify common trends for multiple activities, potentially traceable to the same cause
- Determine candidates for E-Business implementation

Use the review to prompt action. The presentation of positive results gives management the opportunity to be proactive without risk. This proactive attitude is desirable in E-Business implementation.

Begin by assessing whether anyone is doing the work in the activity in terms of measurement and improvement. Lack of activity has both positive and negative sides. On the positive side, it may mean that the new E-Business activities are fine and that people are using them as intended. However, it can mean that the activities have already been changed by the people who operate and use them. These changes could have gone undetected up to now.

Did anyone determine what changes have occurred since the new E-Business activities were implemented? Has the volume increased? Has the mixture of work changed? Has the scope of work changed? Are there more exceptions? Have policies changed?

E-BUSINESS EXAMPLES

RICKER CATALOGS

Ricker faced continuing problems in that marketing generated discounts and promotions that were not conceived of when the E-Business project started. Ricker had a choice—either delay the marketing requests or push through system changes to accommodate the demands of the marketplace. IT devoted all of its resources to the problem and brought in contractor programmers as well.

MARATHON MANUFACTURING

Marathon started E-Business small. They did not face the challenges faced by Ricker. Marathon decided to focus on getting the lessons learned and the wizard in place as well as having the sales representatives visit small firms. This also encountered resistance from the marketing staff, but was overcome with appropriate incentives.

ABACUS ENERGY

Abacus started working with its suppliers. Some of these had limited experience in on-line systems. Abacus finally held seminars and provided no-cost assistance to some of its key suppliers.

CRAWFORD BANK

Crawford found that many customers had questions about loans that were not available on-line. The bank decided to expand its on-line help and instructions on the web. It also found that the web site was not sufficiently user friendly. Some parts of the web site had to be redesigned.

E-BUSINESS LESSONS LEARNED

- **Do not declare victory too early.**
  Wait until you are able to measure the benefits and assess the operational state of the E-Business activities in terms of stability.
- **Make sure that the E-Business activities continue to be measured.**
  When benefits have been determined and interest wanes, you still want to measure the business activities.
- **Adopt the retrofit approach.**
  The retrofit approach is one in which you return to make changes in the business activity that reflect cumulative knowledge and experience in implementation.

WHAT TO DO NEXT

1. Revisit the original issues that were to be addressed by the new business activity. Evaluate these and develop a new list of issues based on the new activity. Place the old issues as rows and the new as columns. In the table enter one of the following codes: N, no relation; C, changed; U, unchanged; I, improved; or E, eliminated.

Issues, new

Issues, old			

2. How many people are involved in each activity in the activity group? List the organizations as rows and the individual activities as columns. Enter the change between the old and the new activity. Run a total at the side and bottom. This will indicate how resources were redeployed.

_____ Business Activities _____ Total

Organization			
Total			

3. Use the data in Action Item 2 to create pie charts for the distributed reduced head count.

4. Create a bar chart in which each organization has two bars (one for old activity and the other for the new one). The height of the bar is the number of people.

5. Return to the tables and charts of Actions 5 and 7 to develop new charts and tables for what really happened after implementation.

6. Develop a table that gives side-by-side characteristics of the old and new business activity.

Business Activity	Volume	Error Rate	Availability	Response Time

7. Consider the following categories of lessons learned: P, project management approach; M, management review and control; I, infrastructure; T, technology; O, organization; S, staffing; V, vendors; I, issue handling; SY, systems; and A, architecture. For each create specific lessons using the following data elements:

   Type: _____

   Title: _____

   What happened: _____

   What should have been done:

   _____

   _____

   _____

   What should be done in the future: _____

What would be the benefit of future actions?

_____

Implementation suggestions:

_____

_____

Part VI

Address E-Business Issues

Chapter 16

E-Business Implementation and Operations Outsourcing

INTRODUCTION

E-Business outsourcing is the transfer of all or part of the E-Business implementation to outside consultants and contractors. It is not restricted to IT activities, but can include business activities such as warehousing, shipping, customer service, order processing, and accounting. Outsourcing was a controversial subject before E-Business and raises many issues related to roles of internal and external groups, accountability, cost, control, and transfer of knowledge—just to mention a few concerns. Outsourcing is a vital part of E-Business implementation.

Examples of areas that can be outsourced for E-Business are as follows:

- Creation of the web site and software customization
- Implementation of the network, hardware, and system software for E-Business
- Market analysis and competitive assessment
- Consulting for E-Business implementation
- Operations of the E-Business systems
- Portals and stores to reduce the cost of the startup costs
- Shipping (through Federal Express or other delivery services)
- Warehousing of inventory
- Processing of returned merchandise, cancellations, and back orders
- Customer service
- Marketing of the web site

The list is endless. If you name any activity of the business, it probably can be outsourced. In IT, outsourcing started in the late 1960s and early 1970s with facilities management and contract programming. Facilities management entails a vendor coming in and taking over the computer systems operation and computer centers. The vendors have more experience and can handle the task economically. Outsourcing is more general than facilities management in that additional services are provided.

E-Business outsourcing can be viewed as a strategic partnership. Both sides bring strengths to the table. They share chemistry and similar cultural values. Recent E-Business outsourcing contracts bear this out. An example is a system integrator who assumes full responsibility for the entire implementation and can achieve economies of scale and bring to bear a wide range of resources to the client firm. Another example is a software firm that develops customized e-commerce systems for firms. The software firm can then use the expertise and software developed with other firms. The effect is cumulative.

Outsourcing makes a company the hub of a network in which vendors are the spokes of the wheel. Firms are increasingly viewing themselves this way. If the network prospers, the company will prosper—if outsourcing is done well. In E-Business outsourcing there are typically several vendors involved to support different aspects of E-Business implementation and operation.

However, outsourcing for E-Business involves risks since you are indirectly putting the outsourcing firm in the network with the customers and suppliers. That is why an entire chapter is devoted to E-Business outsourcing.

DIMENSIONS OF OUTSOURCING

E-Business outsourcing can be thought of having six dimensions. This is a useful way to think of outsourcing overall.

- Relationship of the business activity to the outsourcing issue (why)
- Extent of E-Business outsourcing of the activity (what)
- Type of outsourcing vendor (who)
- Approach that outsourcing firm uses to do the work (how)
- Duration of outsourcing agreement (how long)
- Benefits of outsourcing to the firm (so what)

WHY FIRMS OUTSOURCE

Following is a list of reasons a company would want to outsource E-Business implementation and operation activities. Firms outsource for a combination of these reasons.

- **The E-Business web site is up and running faster.**
  Outsourcing parts of the E-Business implementation makes sense. You have

to be careful in deciding whether to outsource customized software development since the vendor may use the same software elsewhere.

- **The web site shows greater creativity and has higher quality and reliability.**
  A software firm can bring to bear its design creativity and experience. They also are reusing some of their software that is fully tested. This adds to quality and reliability.
- **You lack expertise in E-Business setup and operations, and so outsourcing makes sense.**
  You may not have the experienced staff to handle the network or software support. These are generic IT activities so firms do not view these as core to the business.
- **The current IT staff is tied up in other projects and work.**
  IT groups may be involved in extensive maintenance, enhancement, and software implementation and so cannot spare the resources without a major negative impact on the business.
- **A firm is currently performing a business function such as warehousing for other firms and so they can do it for you at a lower cost.**
  Warehousing, shipping, and other activities may not be viewed as critical to the business. Also, outsourcing activities such as these to established firms can get your E-Business up faster.

Here are some general reasons to outsource E-Business activities.

- **There may be significant cost savings if the outsourcer uses less people to do the work.**
  Staff salaries and benefits are the largest expense component of most companies. E-Business has significant peaks and troughs. Many activities do not involve critical corporate knowledge or expertise. The savings are real and no expertise is lost.
- **Outsourcing is normally a long-term relationship so the contract brings predictability and stability to costs.**
  E-Business costs can get out of control fast. Outsourcing can provide cost stability because a vendor has a legal agreement to provide a specific service at a certain cost for an extended period.
- **Using vendor equipment and facilities can result in major cash flow savings.**
  An outsourcing vendor may promise immediate savings. Alternatively, the outsourcer may purchase the capital equipment from the company. Examples include computer equipment, facilities, and other types of equipment. These generate positive cash flow for the company. Many companies want to use this cash to implement other aspects of E-Business faster.

- **Vendors can offer increased efficiency and effectiveness.**
  Vendors may be more customer focused than internal employees. This makes them more effective. They have performed the same functions for many firms so they are more efficient.
- **Possible tax advantages may accrue through outsourcing.**
  The extent of this depends on the specific firm and situation. There can be tax savings through expensing items that would normally be depreciated over time. In addition, hiring a certain type of vendor in a disadvantaged area may provide additional benefits.
- **Outsourcing offers a way to get skilled E-Business workers and managers.**
  Not only are IT workers with E-Business expertise in demand, but also general managers with E-Business track records. It will take you much longer to get internal staff up to speed. Furthermore, they will be "cutting their teeth" on your E-Business project.
- **Companies see the need to focus on core competencies related to E-Business such as marketing, data analysis, and other company-specific activities.**
  If management and staff do not have to be involved with the outsourced activity, they can turn their attention to more important functions.
- **E-Business may represent a target of opportunity to a firm.**
  An outsourcing vendor may make an irresistible proposal to a firm. Even though the company has not formally considered E-Business, the terms may be too attractive to pass up.
- **E-Business has peaks and troughs of activity depending on seasons, promotions, and other factors.**
  A company may have difficulty rapidly hiring or downsizing to meet demands. If business fluctuations are significant, outsourcing is a way to address the variation. The outsourcer may be providing the same service to many companies, so peaks and troughs in one firm are offset by troughs and peaks in others.
- **E-Business may require that the firm establish a new business activity, such as data analysis or credit card processing.**
  If you begin a new venture or move into a new area, you must acquire expertise quickly. One way to do this is to outsource initial support and performance. As business grows, you can bring these functions back.
- **Your firm may be reaching out to new customers with E-Business, and if your staff is not customer friendly, outsourcing may offer improved customer service.**
  If a department has not been able to achieve adequate customer service, an outsourcing vendor that is service-oriented and whose service performance can measured may be a good alternative.

- **Carried out properly, outsourcing may offer improved management control and accountability.**
  An outsourcing relationship may offer greater control over operations. There are specific standards of performance that can be mandated and enforced that would not be possible internally.
- **Carrying out E-Business support activities can distract management.**
  Manager and employee time may be diverted into a support activity. An example is noncomputer people attempting to fix computer problems or employees trying to manage a new business activity. As a result, major work is being neglected.
- **Outsourcing routine business functions can free up internal resources to focus on E-Business.**
  Some firms outsource software maintenance and operations support so that internal staff can focus on E-Business. This applies to regular business as well.

E-Business Outsourcing Example: A Computer Manufacturer and Its Shipping/Receiving Function

A computer manufacturer outsourced all of its shipping functions to a logistics company in order to concentrate on core E-Business and manufacturing activities. For the firm shipping was not a central activity. The manufacturer's intent was to improve product distribution, control shipping costs, and handle growth. They expected and got multimillion dollar savings and accommodation of growth without increased staff. Previously, the company had projected hiring more than 200 employees to handle growth.

The scope of the outsourcing agreement included all inbound and outbound shipping and domestic and foreign service and repair transportation. The outsourcing vendor was to use its own trucking operation, software, management expertise, and formal business activities. The manufacturer expected to make money by imposing the workload on top of its existing workload and by hiring only a few workers. Outsourcing an entire business function in this manner is becoming more common.

What Business or IT Activity Should You Outsource?

Which business activities should be considered for outsourcing? There are extreme ranges—from a specific function such as planning to the entire implementation of E-Business. You may only outsource an activity for a temporary period, or it can be for years. If E-Business outsourcing is successful, you may outsource more. If it does not, you might bring back these functions inside or hire someone else.

Factors that affect the decision of the activity to outsource include the following:

- Relationship to critical standard business activities. This has to do with competitive advantage and maintaining control.
- Relationship to your long-term E-Business strategy.
- Impact on the speed of the E-Business implementation.
- Impact on the long-term E-Business operations.
- The feasibility of splitting the activity out from everything else and managing it. If you outsource a specific activity in the middle of transactions, problems could arise that bring the business activity to its knees.
- The availability of someone to manage the transition and operation of the outsource vendor. Take the broadest possible view at the beginning of E-Business outsourcing. Then you can narrow the scope later.

WHO DO YOU OUTSOURCE TO?

There are many alternatives for outsourcing E-Business activities. Some of these are:

- Another division of a company. One example is a division that provides parts or services to other divisions or a subsidiary company. The division may have already implemented E-Business with success.
- Former employees or independent contractors. If a core of people perform a function well, but that function is not a core function, you may be able to encourage those employees to form a company of their own. You can then outsource the work to them. This can save money in employee benefits and achieve some of the benefits cited earlier. These people can take over maintenance to free up your staff.
- Contract labor. You can hire someone on a contract or part-time basis to perform specific tasks such as consulting, analysis, and programming. This is the most common approach for E-Business outsourcing.
- Professional outsourcing vendors. You can contract with a vendor that specializes in the specific outsourcing activity, such as accounting, statistical analysis, advertising on the web, and customer service.
- A system integration company that performs the entire implementation and operation of E-Business. An example of this would be a firm that specializes in shipping.
- Subcontractors. A subcontractor performs a specific set of functions year after year.
- Third-party utilities. When a company uses Electronic Data Interchange (EDI) or electronic commerce, it typically uses a third-party network.

This network acts as an intermediary to control the information transfer between suppliers and customers. The same applies to an Internet Service Provider (ISP).

E-Business Outsourcing Tends to Be a Long-Term Relationship

Outsourcing cannot be taken lightly. Some outsourcing agreements typically extend for three or more years. In order to achieve economies of scale and to recover initial startup outsourcing activities, a vendor will often require multiple-year contracts. Also, the company does not want to change vendors frequently. This is especially the case if the company hires temporary employees who must be trained to perform the work correctly.

Over time, the outsourcing vendor and the company develop a relationship and mutual trust. With a new vendor, the relationship must be established all over again. Replacing the supplier or vendor results in a difficult interim period until a new supplier is found or the organization can perform the function internally.

E-Business Outsourcing Issues

So far we have discussed the positive side of outsourcing . However, there are negative issues as well. Some of the most common are as follows:

- Vendor integrity. Will a vendor's knowledge and experience from working with you in implementing E-Business be later used to the benefit of a competitor?
- Lack of vendor expertise. The vendor is no better at the work than the company's employees. The vendor may have claimed specific expertise. However, the work that is provided indicates otherwise.
- Increased costs. The original cost of the contract often increases because the vendor negotiates add-on work. Costs can balloon out of control.
- Overdependence. Management and staff feel comfortable with the vendor. They rely on vendor staff for critical activities and decision making. By the time this trend is recognized, management and staff are too far along the road to change quickly. The business is impacted.
- Inflexible contract terms. The outsourcing work is going badly, and the organization would like to renegotiate the contract. However, this cannot be done because no one read the fine print.
- Loss of control. The vendor begins to take control of the work and indirectly begins to manage some of the organization's departments.

- Loss of in-house expertise. After the agreement is signed, the organization's in-house expertise starts to disappear. It becomes difficult to manage and change the activity.
- Conflict of interest. The vendor gains expertise and experience by working with the organization. To make more money, the vendor markets this expertise to the organization's competitors. If the contract does not prevent this, the organization has little recourse. The organization may be able to get out of the agreement, but the vendor will still have the arrangement with the competitor.
- Vendor turnover. If there is a change in vendors that perform a specific function for your firm, the transition from one to the other may be difficult. Each firm may have a different style as well as method in doing the work.
- Loss of interest. The vendor may have wished to expand into the outsourcing business that it performs for you. However, after some effort, the vendor is unsuccessful. Your business is no longer in their mainstream of work. They may not devote as much attention to it, even it is profitable.
- Loss of business flexibility. If you are dependent on a vendor for your software for E-Business, they may not be able to respond to new requirements on short notice.

There are some tricks of the trade that some vendors practice. Be on the lookout for these.

- Rotating staff. Some vendors start a project with their best managers and staff. These employees make rapid progress. After the startup period, the vendor substitutes a "second string" staff. If the work is routine, this may not be a problem, but productivity may fall and the quality of work may not be as high.
- Using bait and switch. The vendor staff is extremely helpful. The employees do work beyond that for which the customer is paying. The vendor starts charging for additional services. Soon, the customer is paying the vendor more money than initially budgeted.
- Charging for additional tasks. Every time the customer brings up issues, the customer finds that handling these is not within the scope of work.
- Using proprietary tools and methods to lock the customer into a long-term relationship. The vendor comes in and does the work. The tools used are proprietary. The customer now faces increased obstacles to learning what is going on and taking over the work. This occurs in some data processing centers where the vendor has specific tools to maintain and operate computer programs.

How do outsourcers make money? Here are some sources.

- Multiple-year agreements. Organizations will make money over time by making the function more efficient. A major outsourcing contract involving an entire business function is an example.
- Specific tools and methods. The outsourcing vendor has methods and tools that the organization lacks. Thus, the vendor can do the work in less time with less labor. A professional tree-trimming service is a good example of this.
- Experience and expertise in the business activity. The outsourcer knows the activity and can make it more efficient. The vendor also can cut down the layers of management. Vendors employ experienced technical people who can fix problems quickly.
- Economies of scale. The outsourcer already performs the functions so the organization's workload is just added to the vendor's. The vendor can use the same computers for many customers, for example.
- Expanded scope of work. The vendor sees the contract as a foot in the door. The vendor then consciously expands the work's scope and profit.
- Vendors perform the work with lower-paid employees who receive fewer benefits. Vendors may do the work in a different locality, state, or country where costs and taxes may be lower.

Some Guidelines for Considering E-Business Outsourcing

Before you consider E-Business outsourcing, develop a vision of your E-Business future. What will your organization do with respect to E-Business? What will be important to be retained for internal staff? Any outsourcing consideration then relates to this vision. Viewing the company and its vendors as a network is an example of part of a vision.

A company must be willing to deal with greater complexity if it wants to outsource. E-Business outsourcing increases complexity because the company loses total control. There is more coordination with multiple vendors. The company changes from being integrated to becoming distributed, and distributed things are more complex. The trade-off is that by assuming this complexity, the company can grow in terms of service and revenue. Outsourcing encourages a company to develop a more strategic focus. Doing the detailed work internally directs your attention to tactical function rather than the overall picture.

Although outsourcing entails steps similar to those in other chapters, it is unique due to political sensitivities and the involvement of outsiders in your own critical business activities. This is why you should take a conservative position on outsourcing.

The first purpose of this chapter is to determine whether outsourcing will bring sufficient benefits to outweigh the costs, issues, and management of the outsourcing. Beyond that, you want to ensure that the transition to an outsourcing firm is accomplished with minimal disruption. You also want to ensure that the relationship is controlled.

MILESTONES

The major end products include the following:

- Identification of potential opportunities for E-Business outsourcing
- Evaluation and prioritization of outsourcing candidates
- Evaluation of specific activities and recommendations
- Generation of a project plan for outsourcing implementation
- Completion of implementation
- Measurement of the results of outsourcing

METHODS FOR E-BUSINESS

STEP 1: EVALUATE THE FEASIBILITY OF E-BUSINESS OUTSOURCING AND IDENTIFY POTENTIAL FIRMS

To show where E-Business outsourcing would benefit the company, consider your key business activities and their support areas. Some factors to consider are as follows:

- Business volatility. How much will the E-Business change? Are the organization and infrastructure going to be stable? If they are not, outsourcing could lock in obsolete technology.
- Competitive position. Would outsourcing help or hurt the organization's competitive position in E-Business? Would it remove key advantages and knowledge from the achievement of your E-Business strategy?
- Environment in the industry. What are other firms in the industry and similar industries doing? What experience do they have?
- Current state of the business activity. Is the work ready for outsourcing? Is it efficient?
- Retention of intellectual property. Does the outsourcing involve the transfer of intellectual property to the outsourcing firm?

- Alliance benefits. Are there true benefits to the future of the company from an outsourcing agreement that appears to be a joint venture?

Once you have answered these questions, you can identify a set of business activities to be outsourced. There are several general approaches for E-Business.

- Generic E-Business activities
- Specific business activities to free up staff for E-Business
- Specific E-Business activities such as marketing and programming
- General E-Business implementation
- E-Business operation

Next, you can develop a list of potential outsourcing candidates using the same literature search in Action 3.

STEP 2: DEVELOP AN E-BUSINESS OUTSOURCING PLAN

As with E-Business implementation itself, you must develop a plan. You really need this since you will have to figure out how to manage and direct multiple vendors. From experience, we suggest that you assemble an outsourcing team. This will include some of the E-Business implementation project managers as well as other managers from line and support organizations. Select managers who will eventually manage the outsourcing activities. You should also include an accountant and a lawyer. The team members will identify alternatives and determine an initial set of risks and benefits.

The project will start with the investigation and go through to the transition of work to the outsourcing firms. All outsourcing for E-Business should be performed at the same time. This will reduce the overall effort and provide for better cross-vendor coordination as the vendors start their projects at the same time.

The outsourcing plan should include the following:

- The specific business area to be outsourced in regular business and E-Business.
- Interfaces within the business. Address how this area interfaces with other areas of the business. These interfaces may cause problems later if not evaluated during this early stage.
- Interfaces with customers and suppliers.
- The benefits of outsourcing. Identify these, along with the potential drawbacks of outsourcing.
- A backup approach. Assuming that the outsourcing is in place, there must be a backup approach if the outsourcing should fail.
- Measurement and control. Identify the resources required for management of the vendor and how the outsourcing will be managed. The cost of these

resources can be substantial (5 to 10 percent of the total outsourcing cost), diluting financial savings. Spell out how the vendor performance will be measured. Use a vendor scorecard, which may include the following:
— Number of vendor staff involved
— Turnover of vendor staff
— Problems and outstanding issues—the quantity, severity, and age of the problems
— Volume of work performed
— Range of work being addressed
— Response time in doing the work
— Work quality
— Cost of the outsourcing, as well as cost of management of the outsourcing effort
— Benefits of outsourcing to this vendor
• Expanded work policy.
• Changes in business activity needed prior to outsourcing.
• Transition steps.

STEP 3: ASSESS THE BUSINESS ACTIVITIES TO BE OUTSOURCED

Evaluate and analyze in detail any existing business activities being considered for outsourcing. For new areas and activities that have not been established, consider talking to others in the industry as well as consultants. The work here is similar to the activities covered in Actions 5 and 6. The analysis will provide a list of changes that might be implemented in the short term to prepare the activity for outsourcing. If the activity requires nondeferrable major work, remove it from consideration. Analysis should also identify the key people involved. Who will manage the outsourcing vendor? Who will give them additional work as well as monitoring performance? Analyze the economics of the activity, including the capital and operating costs and the tax implications of outsourcing. Treat this work as preparation for disposing of an asset.

STEP 4: IDENTIFY POTENTIAL OUTSOURCING FIRMS

The range of the activities to be outsourced will determine the number and type of firms that you will consider. Since you will be hiring several firms for different activities, the complexity level will rise. They will be interacting with each other. You must manage each vendor, as well as conflicts and problems between vendors. Include vendors who have relevant experience in outsourcing work. Also consider the size of the vendor and its expertise. Your goal is to come up with a vendor list of three to five firms.

STEP 5: GET THE ACTIVITY READY FOR OUTSOURCING

This step applies to existing business activities that you are going to outsource. Implement the changes to these business activities that you identified in Step 3. People sometimes elect to stop the outsourcing plan here if they find that they have improved the activity to the extent that outsourcing benefits are no longer sizable. By improving the activity at this point, you also measure it. This establishes the measurement method for outsourcing later.

STEP 6: OBTAIN OUTSOURCING BIDS

The request for proposal for vendors should contain the following:

- Statement of what the vendor must provide
- What other related areas are to be outsourced to support E-Business
- Additional work that may be requested
- The management method for proposal evaluation and award
- The pricing of services and goods to be provided
- The contract terms of the outsourcing agreement
- Approaches to measurement, management, and dispute resolution
- A standard outline for vendor proposals

To retain control, develop your own contract. If you accept and slightly modify the vendor contract, you may be at a disadvantage. Show the potential firms the current business activity, ideally without disrupting current work. Hold a bidder's conference at which you answer questions. Make clear to firms the key parts of the outsourcing plan from the previous actions, especially the benefits expected, how the firm will be managed, and the approach for controlling changes and additions.

STEP 7: CONDUCT OUTSOURCING NEGOTIATIONS AND SEAL THE CONTRACT

The project team will conduct the evaluation, which typically includes vendor presentations, site visits, or contacts with vendor customers, and in-depth questioning of vendor staff.

Contract negotiations for E-Business outsourcing often focus on these issues:

- Scope of responsibility and services
- Project management
- Cross-vendor coordination

- Skills and abilities to be provided
- Availability of staff
- Warranties
- Rights to proprietary information and indemnification
- Approach to handling change and control
- Dispute resolution
- Measurement method
- Penalties for failure of performance
- Agreement cancellation and termination
- Protection of secrets
- What other customers they can serve and what services they can provide
- Review approach
- Technology upgrades over the life of the contract

Specific business and systems activities may require additional terms.

STEP 8: IMPLEMENT E-BUSINESS OUTSOURCING

Once you have selected a vendor, you will begin the transition or establishment of the activities. Both the vendor and the company appoint transition teams. This is true even for new E-Business activities. You want to define roles and responsibilities for the work. Here are some additional suggestions:

- Use a common project plan so that communications will be easier
- Identify the boundaries of the project and specifically indicate what is off limits
- Work from the same issues database and list to resolve any problems
- Have the vendor staff work with your staff on a substantial percentage of the overall work
- Have weekly meetings to review status, progress, and issues

There are several transition alternatives. You might simply transition at a specific date. You can establish a learning period during which the vendor acquires detailed knowledge and the relationship with the vendor is finalized. A third alternative is to divide the activity into units and then transition each unit.

STEP 9: MANAGE THE OUTSOURCING RELATIONSHIP

Identify a company manager as the chief interface with the vendor. Let the vendor know that no additional work or tasks can be performed without this person's approval. No one else in the company has jurisdiction. The manager will develop measurements for the outsourcing vendor's performance using the vendor scorecard. He or she should have regular vendor meetings to review issues, problems, opportunities, costs, and performance. The manager will track any open items and issues and will give upper management a scorecard of the vendor's performance.

STEP 10: END THE RELATIONSHIP, IF NECESSARY

As with the transition to outsourcing plan, the vendor and company need to assemble teams to carry out a transition back to the company. It is helpful if both sides identify people who have not been involved in problems and issues.

The transition addresses turning over not only the business activity, but also the infrastructure. Transition involves knowledge and training. Vendor staff may have to be hired as consultants to deal with technical issues.

FEEDBACK TO MANAGEMENT

Provide management with feedback during each of the outsourcing actions discussed earlier. The focus of management at all times is likely to be on answers to the following questions:

- Is outsourcing still proving to be economically attractive?
- Did you select the right outsourcing vendor?
- What are the impacts on the business activities?
- What changes have occurred in the activity since outsourcing began?
- Has outsourcing had any impact on employee morale and job satisfaction?
- Are you planning the transition and managing of the outsourcing efficiently?

E-BUSINESS EXAMPLES

RICKER CATALOGS

Ricker implemented activity improvement and design of the new E-Business activities. They also outsourced the shipping function. For web development they acquired a small firm to do the web development. They also hired a firm to get them started in analyzing the marketing and sales data.

MARATHON MANUFACTURING

Marathon contracted out for the setup of the hardware and network. They also contracted with a human resource firm to come up with the incentive program for the sales representatives. However, the development and implementation of the E-Business activities were done internally.

ABACUS ENERGY

Abacus carried out all of the work internally. This was possible since the project was not under extreme time pressure and their supplier relationships were functioning.

CRAWFORD BANK

Crawford lacked expertise in web design and site setup. They contracted with a system integrator firm who performed the hardware and network setup, software development and implementation, and ongoing operations support. The internal activity design was carried out with employees.

E-BUSINESS LESSONS LEARNED

- **Have alternate and backup vendors available.**
  This may sound expensive, but for some activities it is possible. The backup vendor might also be called in to evaluate the work of the primary vendor.
- **Consider dividing the work between several outsourcing vendors.**
  The work, of course, must be divisible. How you design and divide the work to be outsourced determines whether you can use more than one vendor.
- **Reevaluate the outsourcing vendor about halfway through the contract.**
  Do not wait until the end of the contract to evaluate the vendor. If the vendor is not performing as expected, put pressure on the vendor. You can also begin the action steps needed to bring in another vendor.
- **Have a redeployment plan for key people and announce the change prior to the outsourcing vendor selection.**
  Key people should be kept informed to some extent. More importantly, provide information that will make them feel more secure with regard to their jobs after the transition.

WHAT TO DO NEXT

1. What is the track record of your firm relative to outsourcing? More specifically, answer the following questions:

- What is the extent of outsourcing?
- What is the approach for the decision to outsource?
- Is there an effort to gather information on outsourcing?

2. Make a list of functions that link or group together for outsourcing. Include some that are not being considered for outsourcing. Use the list for both rows and columns. Rate the closeness of fit of the functions using a scale of 1 to 5 (1 is no relation; 5 is a tight relationship that would outsource together). Note that a function can include multiple activities.

⟨Function⟩

Function			

3. What interfaces do the functions have with other activities in the business? Develop a table which identifies the nature of the interfaces as well as issues and potential problems.

Function	Interface	Potential Issues

4. Rank the functions in terms of outsourcing. Use the list of functions developed in Action Item 1 as columns in a table. For rows, use various criteria for assessing outsourcing. This will help you see which are the most appropriate functions for outsourcing. By combining the analysis of Action Items 1 and 2, you can combine functions of different suitability for outsourcing. Use a scale of 1 to 5 to rate the functions (1 indicates a major problem with the function for that criteria; 5 indicates that there is a natural solution).

⟨Function⟩

Criteria		
Vendor availability		
Function ready		
Measurement of function completed		
Dependence on company knowledge		
Degree of risk		
Flexibility for E-Business		
Ongoing dependence?		
E-Business technology used		
E-Business experience		

5. Develop a list of 10 to 20 vendors and list them as rows in a table. List the functions as columns. Rate the vendors using a scale of 1 to 5 (1 indicates that the vendor is not suitable; 5 indicates that the vendor is suitable). This table will help you to narrow the number of vendors.

⟨Function⟩

Vendor			

6. What detailed work is necessary to clean up specific functions? Using a table, set the rows as specific actions needed and the columns as functions. Rate the items on a scale of 1 to 5 (1 indicates little or no effort needed; 5 indicates a great deal of effort needed) to determine the relative degree of effort or problem involved. This table will indicate where the cleanup effort should focus.

Function

Organization		
Cleanup steps vs. function		
Training of staff		
Prepare formal procedures		
Develop measurement method		
Streamline current work		
Cleanup information		
Construct new interfaces		
Establish control method		
Downsize		

7. Each function that might be outsourced has its own interfaces between the vendor and the customer organization. The more complex the interface, the more potential problems there are in outsourcing. Use a table to help you assess complexity. The rows are elements of the interface; the columns are the functions. Rate the entry using a scale of 1 to 5 (1 is no problem or issue; 5 signifies a major problem). If many rows are rated 4 or 5, the function will be very complex. Use the following factors for the interface.

- Business activity split with outsourcer
- Internal activities depend on outsourced ones
- Quality control is major concern
- Outsourced function has exceptions
- Sharing of information and extent of information varies over time
- Interface occurs at different frequencies
- Outsourcer must interface systems

Chapter 17

Address Specific E-Business Implementation Issues

INTRODUCTION

Based on our experience in past and current E-Business implementations, we have encountered some of the same issues again and again. In this chapter we deal with 25 of the most common problems that you will likely encounter. These are in five areas: technology, management, business activity, vendor, and organization. For each issue, there is a discussion of how the issue arises and what to do about it when it occurs. Unfortunately, for most of these there is no rule of prevention. Some are almost certainly going to hit you. We want you to be prepared and to plan that some of these will occur in your E-Business implementation.

TECHNOLOGY ISSUES

THE EXISTING TECHNOLOGY IS NOT COMPATIBLE WITH E-BUSINESS

- **How this issue arises**
  The IT group may not fully understand the implications of E-Business support and technology. E-Business requires integrated systems and data. IT may opt for a quick fix. This was true for all four of the examples. Ricker had to implement new hardware and network components and install new software. Abacus piggybacked on their existing systems and technology.

- **Potential actions to take**
  Have the IT group present the overall architecture for both traditional and E-Business support. Ask questions related to change and growth. Look at the architecture from the point of view of a customer or supplier.

THE EXISTING SYSTEMS ARE NOT COMPATIBLE WITH E-BUSINESS

- **How this issue arises**
  The existing systems may have been implemented a decade ago—long before E-Business. The systems were typically implemented for one department. Abacus found that they had to redo their systems related to purchasing and contracting. Marathon was able to use their systems and build on them.
- **Potential actions to take**
  This should be recognized from the start. The firm's management should continuously give this attention. You see on-line compatibility in most cases.

NEW SOFTWARE OR ANOTHER TECHNOLOGY BECOMES AVAILABLE DURING IMPLEMENTATION

- **How this issue arises**
  You should anticipate change. New software and hardware products surface all of the time. This includes new services from firms that will offer to handle your web business.
- **Potential actions to take**
  When you define your E-Business systems and technology architecture, you should consider which areas are weak in terms of products and where new technology is likely to appear.

THE SYSTEMS ARE NOT SUFFICIENTLY FLEXIBLE TO HANDLE THE DEMANDS OF THE MARKETPLACE

- **How this issue arises**
  You implemented E-Business initially. You find that the systems are not sufficiently robust to handle the volume of web activity. You may also find that marketing requirements related to discounts and promotions are not easily handled in the systems.

- **Potential actions to take**
  At the start you should consider a wide range of possible promotions of products. These include across the board discounts, discounts on specific items, and frequent shopper discounts.

New Technology Appears That Makes Some of Your Current E-Commerce Software Obsolete

- **How this issue arises**
  In the item earlier the new software appeared during implementation. Here it appears following implementation. This will naturally arise due to the dynamic nature of the industry.
- **Potential actions to take**
  Have the E-Business coordinators monitor new software along with IT. Implement a regular technology assessment method to see if you should adopt it.

MANAGEMENT ISSUES

Management Changes Its E-Business Strategy in the Middle of Implementation

- **How this issue arises**
  Management may be tempted to make changes based on what they hear in the marketplace. Almost all media are focused on e-commerce and E-Business. As you saw for Ricker Catalogs, changes in strategy are not uncommon based on new information and a greater understanding by management.
- **Potential actions to take**
  You should anticipate this at the start and work with management on a method to assess new ideas and impacts. Changing E-Business strategy can have dramatic impacts on what and how implementation occurs.

Middle-Level Management Resists E-Business while Upper Management Supports E-Business

- **How this issue arises**
  Some middle-level managers may resist change in general. This is also true of senior staff that we have labeled "king bees" and "queen bees." The

bees have power based on knowledge. The term "bees" is used because junior staff run to them for instructions on how to do the work. The bees make it difficult to implement E-Business because they resist participating in E-Business implementation. Ricker had the biggest bee problem and got stung because some of these people did not cooperate.

- **Potential actions to take**
  This book has focused on changing the current activities as well as implementing E-Business activities. The approach to take with middle-level managers and the bees is to focus on E-Business. This is less threatening than directly challenging the current work.

MANAGEMENT PLACES NEW DEMANDS ON E-BUSINESS DURING IMPLEMENTATION

- **How this issue arises**
  This is not the same as changing direction. Here management may want to speed up the implementation. Managers may also want to add transactions to the E-Business transaction list. Management at Abacus decided to implement contracting as an additional application for E-Business. Ricker considered doing customer service, but backed off when they realized the effort.
- **Potential actions to take**
  The best approach is to involve management in implementation issues related to E-Business. By doing this, management starts to see and understand the implications of decisions and actions related to E-Business. This is much more effective than saying "no" when each request surfaces. Another step is to present this as a potential issue at the start of the project. You can also consider different alternatives for the scope of the E-Business project.

MANAGEMENT INSISTS ON MAJOR NEW FEATURES AND CAPABILITIES AFTER INITIAL IMPLEMENTATION

- **How this issue arises**
  Now with E-Business implemented, management may want to insist on more features. This is especially true in E-Business where management and the organization have held back during implementation.
- **Potential actions to take**
  Make the discussion of new features as an activity in your E-Business implementation plan prior to going live with E-Business. This allows you to address the direction of E-Business early and more proactively.

MARKETING OF THE E-BUSINESS SITE IS NOT IN SYNCHRONIZATION WITH IMPLEMENTATION

- **How this issue arises**
  Marketing may either be ahead or behind in implementation. Marketing can get ahead if they begin to generate advertising and promotions before the site is ready. This is natural for marketing to do since there may be a long lead time for print and television media advertising. On the other hand, marketing may be too involved in their normal work. They have no one to assign to E-Business. As a result, they fall behind the implementation.
- **Potential actions to take**
  How do you approach both of these? Our major suggestion is to have a marketing plan that tracks what marketing is doing. The critical advertising dates can be identified early so that the E-Business implementation team can be aware of these dates.

BUSINESS ACTIVITY ISSUES

SOME OF THE ACTIVITIES IMPORTANT TO E-BUSINESS CANNOT BE CHANGED IN TIME FOR E-BUSINESS IMPLEMENTATION

- **How this issue arises**
  An example is when there are separate systems for ordering and customer service. You can replace the ordering activities, but time may not permit the implementation of a new E-Business customer service system. This creates problems since E-Business, when implemented, will be incomplete. Customers may then become frustrated in dealing with manual or partially automated customer service.
- **Potential actions to take**
  You may have no choice but to accept that this situation will occur. Therefore, it is important to plan for this in terms of the analysis of business activities.

AFTER THE IMPLEMENTATION, CUSTOMERS OR SUPPLIERS MAKE NEW DEMANDS ON THE BUSINESS IN TERMS OF VOLUME AND THE NATURE OF THE TRANSACTIONS

- **How this issue arises**
  If your web site is successful, then you will have a problem with addressing the peak demand load on the systems and activities. When you consider this

issue with the previous one, you can see how the customer service area
might be overwhelmed by the new E-Business customers.
- **Potential actions to take**
  Early on you should perform analysis in terms of workload and volume of
  work. You can minimize the impact by staffing planning for transaction
  volume.

Exception Transactions Are Still Being Performed and There Are Still Shadow Systems in Place Despite All of the Effort to Eliminate Them

- **How this issue arises**
  This can typically occur in activities that are changed to handle E-Business.
  You may find that the new E-Business activities cannot handle specific
  types of promotions and discounts, for example. Then the department may
  have to invent workarounds immediately. As time goes on and these are not
  addressed, the department may formalize these internally into shadow
  systems.
- **Potential actions to take**
  The major solution here is to monitor the departments involved in E-
  Business activities or those touched by E-Business. You will want to follow through how specific transactions are addressed. When you identify
  a workaround, then you should support a temporary shadow system and
  work toward modifying the current software to handle it.

E-Business Becomes the Driver for the Business, Causing Structural Problems in the Standard Work

- **How this issue arises**
  Let us suppose that you have implemented E-Business for catalogs and ordering. However, you have made limited changes in shipping and fulfillment. Suppose that the marketing people have implemented special promotions regarding some items in terms of free shipping and handling. When
  the promotion hits, the shipping department and its systems cannot handle
  this promotion. They were never designed for this. The impact may be a
  surge in customer complaints that require substantial manual handling, de-
  stroying the efficiency gains of E-Business in ordering.
- **Potential actions to take**
  You should plan ahead for where the potential bottlenecks will appear in the

standard work. You can assume that any existing business activity that will be touched by E-Business transactions will be hit.

E-BUSINESS GROWS FAST, CREATING TOO MANY TRANSACTIONS FOR THE TRADITIONAL BUSINESS ACTIVITIES

- **How this issue arises**
  When you have implemented E-Business, you do not really know how much business you will get and how fast it will grow. Most people err on the conservative side, believing that they can make up for any problems later. For automated tasks it may be possible to add more hardware. However, this is not easy for the traditional activities that are more labor intensive. The difficulty is that the problems occur in real time.
- **Potential actions to take**
  You should be ready to add staff in some departments quickly. Up-to-date training materials should be ready. These materials should recognize and support E-Business transactions.

VENDOR ISSUES

THERE ARE PROBLEMS IN COORDINATING THE ACTIVITIES OF MULTIPLE VENDORS IN E-BUSINESS IMPLEMENTATION

- **How this issue arises**
  In E-Business you may have several vendors. You could have one firm implementing the e-commerce software. Another firm may be involved in doing system integration. Coordination of the vendor staff is crucial to the success of the project.
- **Potential actions to take**
  Have regular meetings with the vendor managers. Support and insist on sharing common project plans and lists of issues. Assign tasks that must be performed by staff from several vendor firms. These actions support a collaborative approach.

THERE IS A LACK OF QUALIFIED PEOPLE AVAILABLE WHO KNOW E-COMMERCE SOFTWARE

- **How this issue arises**
  This is a general problem with any expanding and new field. It takes time for technical people to first embrace the software and then to gain expertise

with it. The consequence is that you may be short on people so that implementation is delayed. This may also impact support later.

- **Potential actions to take**
  Knowing that this problem is likely to occur helps. You can raise it as an issue and then attempt to address it through training. You also may have sufficient funds to be able to acquire a small software firm. Two of our examples did this. Another action to take is to work toward spreading and disseminating knowledge of the software.

VENDOR PRODUCTS HAVE GAPS IN FEATURES THAT ARE IMPORTANT TO E-BUSINESS IMPLEMENTATION

- **How this issue arises**
  Sometimes, you do not know about a technology gap until you experience it in practice. It is then too late. The technology gap now necessitates a workaround that could have a negative impact on the customer or supplier experience on the web.
- **Potential actions to take**
  To head this off you want to talk with users of the software and see what problems they ran into during implementation. You also need to develop tables that show how the software products and systems will interface with each other.

IT IS DIFFICULT TO GET TIMELY VENDOR SUPPORT TO RESPOND TO PRODUCTION PROBLEMS

- **How this issue arises**
  The vendor may have moved onto other clients. They are trying to develop new software products, perhaps. There is a shortage of talented and experienced people. These factors tend to impact vendor support. The impact may be more substantial in E-Business since some of the problems may be visible to the customers or suppliers using your web site.
- **Potential actions to take**
  You should plan ahead for support needs. This is different from traditional thinking where you consider support in terms of fixes that may take considerable time and delays to implement.

THERE IS A FEAR THAT VENDORS WILL TAKE THEIR EXPERIENCE AND KNOWLEDGE FROM YOUR PROJECT AND GO TO THE COMPETITION—YOUR COMPETITIVE ADVANTAGE MAY JUST EVAPORATE

- **How this issue arises**
  Vendors will take what they learned from their work with one client to the next ones. You really cannot block this. Vendors cannot be forced to stop this.
- **Potential actions to take**
  There are several approaches to take. One is to contract with the vendors on an ongoing basis. This makes sense if you are going to continue expanding E-Business. Another approach is to restrict what vendors do and to limit the vendors in terms of activities. Thus, the internal staff would perform more of the proprietary work.

ORGANIZATION ISSUES

PEOPLE CONTINUE TO VIEW E-BUSINESS AS A THREAT TO THEIR JOBS

- **How this issue arises**
  Some employees see E-Business as moving the business transactions toward the web where the customers and suppliers perform the work. No matter what management and the implementation team say, this feeling can continue.
- **Potential actions to take**
  This condition must be considered as a given. It should be assumed to exist. The major approach is to involve human resources in the method.

MARKETING GETS AHEAD OF IMPLEMENTATION WHEN THEY DEFINED A NEW PROMOTION

- **How this issue arises**
  Marketing can be very enthusiastic toward E-Business. They see it as a challenge. They make contacts with marketing people at other firms and start to devise creative promotions and discounts. Unfortunately, this

information is not shared with the implementation team or with IT. The new E-Business may not be able to support these promotions and discounts.

- **Potential actions to take**
  With a separate marketing plan and discounts and promotions getting active attention from the implementation team, there should be improved communications. This should help head off future problems when E-Business goes live.

As E-Business Gets Going and Takes Off, There Is a Need for Major Ongoing Change in the Organization

- **How this issue arises**
  Most organizations try to implement E-Business either as a separate new business or with their current organization. There is so much to do in E-Business implementation that people do not pay attention to organization. After all, why rock the boat even more? The need for organization change typically manifests itself in several ways. First, some parts of the organization may not be involved in E-Business. It is not that they are resisting—they just cannot cope. A second way is that organization politics and problems affect the E-Business work. For example, customer service might suffer. Customer service then blames the ordering activities. The managers involved in E-Business ordering then blame customer service for not getting on board.
- **Potential actions to take**
  You should recognize that some organization change is inevitable. If so, then should you rush in and define the new organization ahead of time? No. You should plan ahead for a new organization. Keep it quiet until issues arise. The issues will then be a trigger to initiate the organization change.

There Is a Lack of People in the Organization Who Grasp the Implications of E-Business

- **How this issue arises**
  E-Business and e-commerce may be vaguely known in terms of concepts, but there is usually no in-depth understanding. People may think of E-Business in terms of a particular company. The impact of this lack of knowledge is both an opportunity and a potential problem for the E-Business implementation team.

- **Potential actions to take**
  The first action is to understand how people do their business in the traditional ways today. Do not rush in and try to explain how great E-Business is. Staff will feel more uneasy. As you understand the business, you can start to show the employees how E-Business would work through transactions they are familiar with. This will give them a more useful understanding of E-Business. You can then get them involved in ironing out the transactions' details for E-Business.

ORGANIZATIONAL CONFLICT AND DEPARTMENTAL POLITICS DELAY AND NEGATIVELY AFFECT THE E-BUSINESS IMPLEMENTATION

- **How this issue arises**
  E-Business requires departments that have never really worked together to now share information, define common activities, and eliminate and change how they do their work. Many established companies have departments that have evolved into little empires. E-Business challenges these. E-Business also challenges the exceptions, workaround procedures, and shadow systems.
- **Potential actions to take**
  You begin by working to understand the internal power structure. You must take nothing for granted. Even if department managers are openly supportive of E-Business, the supervisors and lower-level managers may resist. Raise this as an issue at the start. Consider how transactions are performed across departments. Do this by observation and by interviewing.

CONCLUSIONS

With all of the challenges to implementing E-Business, you might be inclined to avoid it, or to at least wait until the technology becomes more stable. Unfortunately, time is not on your side. A more "wait and see" traditional approach might work in slower-moving situations, but it fails here. E-Business implementation is different in several major ways from traditional process change and systems implementation. First, there is the business dimension. Marketing and management are heavily involved from the start. Second, E-Business is buffeted by outside forces. Customers, suppliers, the competition, and technology all bear down on your E-Business efforts. Third, E-Business implementation is recurring and ongoing. You do not just implement and walk away. All these factors combine to make E-Business implementation one of the most exciting and rewarding areas of business and technology today.

The Magic Cross Reference

(Continues)

(*Continues*)

Area	Topic	Pages
Outsourcing	Activities for outsourcing	291–292
	Contracting	299–300
	Guidelines	295
	Issues	293–294
	Management	300–301
	Outsourcing plan	297–298
	Reasons for outsourcing	288–289
Technology assessment	Architecture	97–98
	Categories of software	98
	Infrastructure components	103

Index